1700 Main Street, PO Box 2138 Santa Monica

ADDENDUM

PUBLICATIONS DEPARTMENT

March 1, 1999

To: Recipient of MR-960-A

Title: Emerging Commercial Mobile Wireless Technology and Standards: Suitable for the Army?

Author: Phillip M. Feldman

The following paragraphs are an addendum to Section 4.3 ("Waveforms and Signal Processing"), which begins on page 40.

Factors that work against common military and commercial approaches to waveforms and signal processing

There are three main factors that tend to force a break between the (tactical) military and commercial worlds in the area of waveforms and signal processing (not all of these issues are explicitly raised in the report):

1. Commercial waveforms and signal processing do not provide for (a) resistance to jamming, (b) low probability of detection, and (c) operation without a supporting fixed (in-theater) infrastructure.

Item (c) is a major issue for network protocols, but is also important from the point of view of signal processing. Commercial cellular networks use centralized base stations that are carefully sited in order to minimize mutual interference between base stations and between users in different cells. Tactical networks will either not use base stations at all, or (more likely) will use base stations that are mobile or at least transportable. Careful siting of tactical base stations is of course out of the question. The commercial CDMA (code division multiple access) cellular systems use direct sequence spread spectrum with multiuser power control to control interference between users; this power control involves commands from the base station to all handsets in a cell that adjust the transmit power level with a very high update rate. For a variety of technical reasons, this type of power control cannot be used in tactical wireless networks. (This discussion does not apply to other types of power control, e.g., power control to maximize battery life). In short, advanced interference control techniques such as multi-user detection are a much higher priority for the military than for commercial wireless applications.

2. Military procurements are extremely small by commercial standards. A rough rule of thumb is that application-specific integrated circuits (ASICs) are cost effective for equipments that are being manufactured in volumes of 100,000 and above, and are typically not cost effective for volumes less than 10,000. (The exact values of these numbers are of course open to dispute, but the exact numbers are also not critical for this discussion). For many military communications equipments, field programmable gate arrays (FPGAs) and programmable digital signal processors (DSPs) will be more cost-effective than ASICs.

3. Ideally, the military should have equipments that can be reconfigured to (a) take advantage of new waveforms and algorithms, (b) cope with new threats, and (c) meet both changing operational requirements and the requirements of specific missions. This is of course only practical in equipments that use programmable devices.

Some high-end commercial wireless equipments, including IS-95 and GSM digital cellular phones, phones for use with Low Earth Orbit (LEO) satellites, and multimode cellular/satellite phones, are using or will use hybrid ASICs that combine analog, nonprogrammable digital, and programmable DSP elements on a single chip. (LSI Logic's CDMA baseband processor chip, the CBP 2.0, is an example of such a chip.) Use of chips that have programmable elements but are not fully programmable allows for faster product development and increased flexibility with only modest increases in cost. Commercial wireless system designs based on fully programmable signal processing will not happen in the near future. Both factors 2 and 3 tend to push the military toward the use of fully programmable signal processing, whereas commercial equipments that are manufactured in large volumes will generally use ASICs (with possibly some programmable elements) in order to minimize power consumption, size, weight, and cost.

RAND

Emerging Commercial Mobile Wireless Technology and Standards

Suitable for the Army?

Phillip M. Feldman

Prepared for the United States Army

Arroyo Center

Preface

For the study entitled "Fundamental Research Policy for the Digital Battlefield," the main goals were:

- To evaluate commercial wireless communications technology, including products and services; components and subsystems; protocol standards; and waveforms and signal processing techniques, in order to determine their suitability for Army tactical applications, and to suggest the appropriate mix of commercial, military-unique, and military variants of commercial systems for use on the digital battlefield.
- To recommend specific Army 6.1/6.2 research areas where progress is needed to address gaps between military requirements and presently available and emerging technology.

Study results are documented in this report and in Leland Joe and Phillip M. Feldman, *Fundamental Research Policy for the Digital Battlefield,* DB-245-A, forthcoming. The research was sponsored by the Assistant Deputy Chief of Staff for Combat Developments, U.S. Army Training and Doctrine Command (TRADOC), and was conducted in the Force Development and Technology Program of RAND's Arroyo Center. The Arroyo Center is a federally funded research and development center sponsored by the United States Army.

The report should be of interest to government and military decisionmakers, system designers (both military and commercial), research officers (within the Army and in other organizations that fund research), and standards-making bodies. In order to reach the widest possible audience, the report has been written so as to require only a minimal knowledge of communications engineering concepts and terminology.

Contents

Appendix

Figures and Tables

Figures

Tables

Summary

This study, entitled "Fundamental Research Policy for the Digital Battlefield," was sponsored by the Assistant Deputy Chief of Staff for Combat Developments, U.S. Army Training and Doctrine Command (TRADOC). The primary objectives of this report were the following:

1. to evaluate the suitability of commercial wireless technology (products, services, waveforms and protocols, components, and standards) for Army tactical applications,

2. to recommend areas where Army 6.1/6.2 research funding might yield solutions to wireless communications problems that are of particular concern for the military and thus unlikely to be addressed by industry-funded research, and

3. to explain the competing factors (tradeoffs) that must be weighed when selecting between alternative systems or technological solutions, and to suggest a general methodology for understanding these tradeoffs.

Although current-generation commercial and military ground-based mobile wireless networks might seem superficially to be providing similar functionality, they are in fact fundamentally different types of networks; future generations of these ground-based wireless networks may diverge even further. Commercial wireless networks, which are perhaps best exemplified by the cellular telephone systems, depend on a fixed supporting infrastructure of base stations interconnected by high-speed trunk lines. The topology of the supporting network is fixed, which greatly simplifies the routing of connections. Military ground-based mobile wireless networks cannot depend on a static supporting network for several reasons; the most important of these are (1) highly mobile forces need networks that move with them, (2) fixed assets are more vulnerable to attack, and (3) the military needs networks that will continue to function even when some nodes are destroyed and some links are jammed.

Many architectures have been proposed for mobile military networks of the future. These include packet radio networks in which every network node acts as a packet switch, networks in which users communicate though satellites or pseudo-satellites (e.g., on UAVs), hierarchical ground networks in which some nodes have switching and routing capabilities while others do not, and hybrids that combine these concepts in various ways.

Some hierarchical network concepts are essentially modified cellular telephone networks with mobile base stations, retaining the circuit-switched architecture of cellular networks. These networks may be suitable for connection-oriented traffic, but they are inefficient for messages and short data transmissions. The packet radio network concept, on the other hand, was originally developed for transmission of data and short messages, and it is well suited for this type of traffic but not easily adapted to such connection-oriented traffic as real-time voice and video. Messages and short data transmissions are a large and growing fraction of tactical network traffic, but interactive voice and other connection-oriented traffic will remain an essential component of tactical network traffic. Thus, future mobile military networks must include elements of both circuit switching and packet switching in order to support a mix of connection-oriented and connectionless traffic.

Documents such as the Joint Technical Architecture (DoD, 1997) show that the Army is committed to the integration of standard protocols and commercial technology such as TCP/IP and ATM. There are, however, important mismatches between these commercial standards and military requirements. Existing and emerging commercial wireless standards are gradually addressing many of the communications problems that must be solved to meet the needs of the commercial world for wireless voice, wireless e-mail access, and related services. But at both the physical layer and higher layers, choices are being made that are fundamentally incompatible with Army tactical operations and with the Digital Battlefield concept.

The military will need, for example, network backbone mobility, i.e., mobility of both hosts and routers, but current IP standards do not provide for this—neither the Mobile IP extension of IP version 4 nor the mobility component of IP version 6 has any provision for mobile routers. Commercial wireless (terrestrial and space-based) systems and services will not meet the Army's future tactical needs, and the Army must consequently trade off requirements against future investments in research and Army-unique systems.

The picture is less bleak in the components area. For many types of components, commercial and military requirements are fairly close, and the military can either use unmodified commercial components or arrange for the manufacture of military variants of standard commercial components. For some types of components, however, military-specific components will continue to be necessary, and the diminishing pool of suppliers in some of these areas gives cause for alarm. Where there is insufficient commercial demand for a class of component or subsystem that the military needs, it may be worth taking aggressive measures to ensure the continued existence of a reasonable pool of both suppliers and R&D technical expertise.

Mismatches between Army requirements and commercial standards/practices will have to be addressed by changing the requirements, departing from commercial practice, inducing (to the extent possible) necessary changes to and extensions of commercial standards, or (most likely) a combination of all these.

Military systems designers and planners have a critical need for simulation tools that can accurately predict the performance and behavior of mobile wireless networks operating in realistic tactical environments. Existing tools tend to concentrate on either the middle protocol layers or the lower "physical" layer, and they do not simultaneously model all of the layers with sufficient detail and accuracy to yield useful results.

There is a need for (A) models that can accurately assess the impact of mutual interference (both co- and adjacent-channel) when large numbers of equipments operate in close proximity, (B) models that can be used to compare narrowband, frequency-hopping spread spectrum, and direct-sequence spread spectrum systems operating within a mobile network, including multipath effects, and (C) standard channel reference models against which competing system and network concepts can be tested.

Acknowledgments

I would like to thank Dr. Leland Joe of RAND, who directed this study, and Mr. Bill Cunningham of the Army Training and Doctrine Command for helpful comments on early drafts of this document, as well as many stimulating discussions.

Glossary

A/J	Anti-jamming. Any feature of a communications or radar system designed to increase its ability to function properly in the presence of jamming.
AMPS	Advanced Mobile phone system. An analog cellular telephone service available over much of North America.
API	Application programming interface. The set of function calls (procedure calls) via which a computer program accesses a given set of services. An API is specific to a given language, e.g., C, C++, or FORTRAN 77.
ARQ	Automatic Repeat reQuest. A type of error control in which packets that are not acknowledged by the receiver are re-transmitted after a specified time delay. Hybrid ARQ refers to a combination of ARQ with FEC.
ASIC	Application-specific integrated circuit. (As opposed to a general-purpose integrated circuit, such as a memory chip or CPU chip).
ATM	Asynchronous transfer mode. A communications network protocol which is essentially packet switching at the lowest level, but is capable of supporting not only connectionless traffic but also (like circuit switched networks) real-time connection-oriented traffic such as interactive voice and real-time video.
AWGN	Additive white Gaussian noise.
bandwidth	The strict definition of *bandwidth* is the width in Hertz of the frequency spectrum occupied by a given signal, or by some fraction of the signal power (e.g., 99 percent). In common parlance, *bandwidth* has come to be equated with *data rate*, measured in bits per second (bps).
BER	Bit error rate. The long-term average fraction of bits received in error. Same as bit error probability.
CDMA	Code division multiple access. A type of spread spectrum multiple access in which simultaneous transmissions avoid (or minimize) mutual interference by using different spreading codes. CDMA can be implemented with either DS SS or FH SS (commercial CDMA uses DS SS).
COTS	Commercial off-the-shelf. Denotes standard commercial components and equipment, as opposed to items designed specifically for the military market.
DAMA	Demand assignment multiple access. In satellite systems and in other networks with repeaters, a class of schemes for sharing satellite channels among a population of users.
DS SS	Direct-sequence spread spectrum.
ECM	Electronic counter measures.
EIRP	Effective (or equivalent) isotropic radiated power. EIRP is a measure of radiated power density, and is equal to the product of actual radiated power and antenna gain. Units are watts.
EMC	Electromagnetic compatibility. The ability of two or more equipments to operate together (under specified conditions) without causing unacceptable interference to each other.

ESM	Electronic support measures. ESM refers to systems that perform direction finding or that are used for measuring/characterizing the spectral or time-domain characteristics of RF emitters.
FDMA	Frequency division multiple access. A basic waveform type in which a band of frequencies is divided into smaller nonoverlapping sub-bands, or channels. A given transmission uses only one of these channels.
FEC	Forward error control. Error control that, unlike pure ARQ, does not involve retransmissions. FEC detection and/or correction depends on structured redundancy only. The FEC code rate equals the number of information symbols (before encoding) divided by the number of channel symbols (after encoding).
FH SS	Frequency-hop spread spectrum. A given transmission uses a pseudo-random sequence of transmit carrier frequencies.
FIR	Finite impulse response. A type of linear filter. Output at time t is weighted average of inputs over $[t - t, t\,]$.
FLOT	Forward line of troops (formerly, forward edge of battle area, or FEBA).
FPGA	Field programmable gate array.
GPS	The global positioning system is a constellation of 24 satellites that provides navigational information to military and civilian users. Signals from any four satellites enable a user to determine three-space position and time. For users on the surface of the earth, three GPS satellites must be visible. GPS provides two sets of signals; the less accurate *standard positioning service* is available to everyone.
HCTR	High capacity trunk radio.
IETF	Internet engineering task force. A standards-setting body that has de facto control over Internet-related standards, although no legal standing.
IP	The Internet protocol. IP is the network layer protocol in the TCP/UDP/IP Internet protocol suite. Both TCP and UDP depend on IP.
ISDN	Integrated service digital network, a group of digital services providing point-to-point circuit switched connections for voice, data, facsimile, and video at rates ranging from 64 kbps to 1.544 Mbps.
ISO	The International Standards Organization. One of many international standard-setting bodies.
JSI	Jammer side information or jammer status information. JSI is symbol-by-symbol information about the presence or absence of jamming, and can aid the FEC decoder.
JTIDS	The Joint Tactical Information Distribution System, a military radio for broadcast messaging and voice.
LAN	Local area network.
LEO	Low Earth orbit. LEO designates the regime of altitudes between 175 and 2,000 km. Most circular orbits at altitudes between about 2,000 and 10,000 km (1,250 to 6,250 miles) are impractical because the trapped radiation in the inner Van Allen belt causes damage to unshielded electronics. Because of atmospheric drag, altitudes below about 175 km (109 miles) decay too rapidly to be practical.

LPD	Low probability of detection. LPD is often used to refer to both signal detection and direction finding ("localization"), since in practice these are closely-related activities.
MEO	Medium Earth orbit. MEO designates the regime of altitudes above 10,000 km but below geostationary altitude (35,860 km). See LEO.
MIPS	Millions of (integer) instructions per second. A measure of computer performance.
MSE	Mobile subscriber equipment. See also MSRT.
MSRT	Mobile subscriber radio telephone. The AN/VRC-97 mobile subscriber radio telephone (MSRT) is the MSE mobile subscriber terminal.
OSI model	Open systems interconnection model. An ISO standard conceptual model for communications networks that divides functions into 7 layers. In existing networks, functions are not always organized according to the OSI model.
PCS	Personal communications services. A class of digital wireless systems that provide two-way voice in combination with at least one other nonvoice service such as text messaging.
PEM	Plastic encapsulated microcircuit.
PSD	Power spectral density. Watts/Hz as a function of frequency.
PSN	The public switched network. The interconnected network of switches, trunks, digital cross-connect systems, and customer premises equipment that supports leased telephone lines, analog switched services (e.g., telephony and facsimile), and digital switched services (e.g., frame relay and ATM).
QAM	Quadrature amplitude modulation. A modulation in which each symbol represents b bits, and is transmitted as a weighted combination of sine and cosine components at the carrier frequency. Each of the 2^b possible symbols is represented by a different pair of (real-valued) weights.
QoS	Quality of service for connection-oriented services, measured in terms of such parameters as throughput, delay, and delay jitter.
SFH	Slow frequency hopping. See FH SS.
SNR	Signal-to-noise ratio (signal power divided by noise power). SNR is a useful measure of signal quality when any interfering signals can be collectively treated as additive white Gaussian noise over the frequency band of interest.
TCP	Transmission control protocol. TCP is the connection-oriented transport layer protocol in the TCP/UDP/IP Internet protocol suite. See UDP.
TDMA	Time division multiple access. A basic multiaccess waveform type in which time is divided into slots, typically of fixed length; a given transmission must fall entirely within a single slot.
UAV	Unmanned aerial vehicle.
UDP	User datagram protocol. UDP is the connectionless transport layer protocol in the TCP/UDP/IP Internet protocol suite. Unlike TCP, UDP performs "best effort" delivery. UDP does not generate acknowledgments, does not retransmit missing packets, and does not guarantee in-order delivery.
VLSI	Very large scale integration.
WLAN	Wireless local area network.

1. Introduction

1.1 Purpose and Scope of the Study

This study, entitled "Fundamental Research Policy for the Digital Battlefield," was sponsored by the Assistant Deputy Chief of Staff for Combat Developments, U.S. Army Training and Doctrine Command (TRADOC).

The U.S. Army, as well as the other services, is moving in the direction of greater use of commercial technology and standards. The principal motivation for this change is the desire to reduce costs. However, increased interoperability is another potential benefit. The Technical Information Architecture generated by the Army Science Board in the summer of 1994 (ASB, 1994) highlighted the problem of stovepiped systems and the lack of interoperability among military communications systems. The Internet Protocol (IP) was subsequently accepted as an element of the Army and Joint Technical Architectures (DoD, 1997).

A number of recent studies have examined various issues relating to military use of wireless communications technology, products, and standards. Several National Research Council panels have issued relevant reports; these include "The Evolution of Untethered Communications" (NRC1, 1997), "Energy-Efficient Technologies for the Dismounted Soldier" (NRC2, 1997), and "Commercial Multimedia Technologies for Twenty-First Century Army Battlefields" (NRC3, 1995).

The term "standard" is used here in the broadest sense, including not only official *de jure* standards of recognized standards organizations such as the Institute of Electrical and Electronics Engineers (IEEE) and the International Standards Organization (ISO), but also de facto standards which are endorsed only by the marketplace.

1.1.1 Goals of the Study

The main goals of this study were as follows:

1. To evaluate the current state and likely near-term[1] development directions of commercial wireless technology, including both standards and products, and in particular to identify areas where there is a mismatch between the needs of the Army and what the marketplace can be expected to offer without government intervention. Although the term "wireless" includes not only ground wireless communications, but also air-to-ground communications, air-to-air communications, and satellite communications, the emphasis here is on nonsatellite wireless systems suitable for Army tactical communications.

2A. To explain the competing factors (tradeoffs) that must be weighed when selecting between alternative systems or technological solutions, and to suggest in general terms how these factors should be weighed.

---

[1]Because reliable long-term predictions are impossible in a field that is changing so rapidly, we consider only developments that are likely within the next decade.

2B. To suggest where the Army is likely to need military-unique solutions, where purely commercial solutions are likely to be acceptable, and where a mix of the two might be appropriate.

3. To identify research areas where judicious application of research funds could lead to solutions for problems that the commercial world is not moving aggressively to solve, but where "ready made" solutions might find willing adopters.

We have drawn on the research of others wherever suitable research results were available, and extrapolated when such data were not available. A thorough treatment of 2A is outside the scope of this study (and indeed outside the scope of any single study).

A wide variety of problems involving modeling and analysis/performance prediction for wireless systems and networks are currently the subject of active research by many institutions and individuals. Thus, with one exception, we do not treat modeling and analysis issues in this study. The exception is an area that we regard as the largest single roadblock preventing (honest) analysts from making fair comparisons among competing systems; this provides our fourth goal:

4. To explain the channel modeling problem, to motivate the need for a solution or solutions, and to suggest some of the elements that such solutions should have.

Although the above discussion might be understood as suggesting that all of the problems the military faces in integrating commercial wireless technology into its networks are technical in nature, we believe that there are other components of the problem. Availability of funds for modernization and system integration problems come to mind immediately. However, there are also "educational" problems, e.g., most commercial communications systems developers have at best a minimal understanding of what the military needs, while, on the other hand, military decision-makers are sometimes ignorant of commercial trends, market forces, and related issues that should inform their choices.

1.1.2 Organization of the Document

The remainder of this document is organized as follows:

Subsection 1.2 briefly describes some of the more important classes of wireless communications equipments, and presents two alternative classifications of mobile wireless networks. Subsection 1.3 provides some general background on circuit-switching and packet-switching concepts and terminology (readers who already have some background in networking may wish to skip Subsection 1.3).

Chapter 2 explains some of the more important performance measures and design tradeoffs for commercial wireless systems. Chapter 3 presents and compares three alternative architectures for future military mobile wireless networks: rapidly-deployable cellular networks, mobile mesh networks, and fully mobile hierarchical networks. Because of the current high level of interest in this area, we discuss several possible implementations of hierarchical networks based on the use of airborne relays. Because the existing Joint Tactical Information Distribution System (JTIDS) also provides for airborne terminals and airborne relays, we compare these approaches and explain how the hierarchical networks can avoid some of the limitations of the JTIDS full-mesh network architecture.

Chapter 4 evaluates the tactical utility of commercial wireless technology, products, protocol standards, and services. In Chapter 5 we discuss some issues in channel modeling and explain why standard reference channel models are needed. Chapter 6 presents our recommendations for 6.1/6.2 research funding that might address some of the shortfalls and problems addressed in Chapter 4. Conclusions and major findings of the study are stated in Chapter 7.

Four appendices at the end of the report provide background material that may be helpful for some readers: Appendix A gives a brief overview of tactical radio communications. Appendix B gives a brief review of multiple access communications, including a comparison of different spread-spectrum techniques. Appendix C provides general background on channel models. Appendix D derives a form of Claude Shannon's capacity formula that exposes the relationship between spectral efficiency and power efficiency.

1.2 Taxonomies of Wireless Communications Networks

1.2.1 Background

Important types of wireless communications equipments (commercial and military) include:

- cellular phones. These can be analog or digital. Mobile users are supported by a fixed (or in special cases transportable) infrastructure of base stations interconnected by high-speed trunk lines. Base stations support handovers so that users can move from one cell (region supported by a given base station) to another cell.
- cordless phones. For most types of cordless phones, each mobile user is supported by a single base station. A user cannot move from one base station to another while a call is in progress. (The European DECT cordless phone system does have handover capability).
- line-of-sight radios. These are primarily of two types:
  - (A) Fixed or transportable, high capacity systems for point-to-point trunking (multiple streams of data and digital voice are multiplexed together over a single connection). The Army's High Capacity Line of Sight (HCLOS) Radio program, which is intended to "serve as the next generation line-of-sight (LOS) radio for the Warfighter Information Network (WIN)," is of this type. "Its primary role will be interswitch links on the [network] backbone" (Gordon, 1998).
  - (B) Mobile, semi-mobile (stop to transmit or receive), or transportable low-capacity radios that are designed primarily for handling single two-way (typically push-to-talk) voice connections. The AN/GRC-240 HAVE QUICK II ground radio is of this type.

  In both cases, connections are limited to line of sight because of operation at frequencies above HF, the need for high data rates, lack of hardware and protocol support for multihop transmission, or some combination of these factors.
- packet radio networks. The radios in these networks are digital and exchange information in a store-and-forward fashion, so that a source and destination that are not able to communicate directly may nevertheless be able to exchange information. Packets are routed through the network, and may take one or more hops to reach the destination. See Bertsekas and Gallager (1992) for a general overview of the packet radio network concept. The survey

issue of the IEEE Proceedings edited by Leiner et al. (1987) and the article by Kahn et al. (1978), although somewhat dated, are also useful.

- pagers. These include conventional pagers, alphanumeric pagers, and two-way pagers.
- satellite terminals.
- wireless local area networks (WLANs).
- wireless modems.

Wireless communications hardware is often a separate piece of equipment used either in a stand-alone fashion (e.g., cellular telephones, cordless phones, and pagers) or connected to a computer for the purpose of data communications (e.g., wireless modems).[2] In some cases support for wireless communications may be integrated into a piece of equipment that performs other functions as well.

The primary *raison d'être* for wireless communications is to enable mobility, and this is also the principal benefit for the military. However, "mobility" means different things in the commercial and military worlds. Corson and Macker (1996) succinctly explain the differences between what we refer to as "fully mobile wireless networks" and wireless networks supported by a fixed infra-structure:

> Within the Internet community, the current notion of supporting host (user) mobility is via "mobile IP." In the near term, this is a technology to support host "roaming," where a roaming host may be connected through various means to the Internet. However, at no time is a host more than "one hop" (i.e., a wireless link, dial-up line, etc.) from the fixed network. Supporting host mobility requires address management, protocol interoperability enhancements, and the like, but core network functions such as routing still occur within the fixed network.
>
> A long term vision of mobile IP is to support host mobility in wireless networks consisting of mobile routers. Such networks are envisioned to have dynamic, often rapidly changing, mesh topologies consisting of bandwidth-constrained wireless links. These characteristics create a set of underlying assumptions for protocol design which differ from those used for the higher-speed, fixed topology Internet. These assumptions lead to somewhat different solutions for implementing core network functionality such as routing, resource reservation, etc.

In the remainder of this subsection we present two alternative classification schemes for the mobility aspect of communications systems/networks.

1.2.2 A Classification Based on Network Architecture

Communications networks can be divided into four categories based on characteristics of the supporting infrastructure:

(1) Wireless systems with a fixed supporting infrastructure. Most existing wireless systems fall into this category. A mobile user connects to a base station, access point, or satellite gateway; the remainder of the communications path (assuming mobile-to-fixed communications) passes

---

[2]Some digital cellular phones have data ports and thus can be used either stand-alone for voice or connected to a computer for data.

over "wired" networks.[3] Examples include cellular phone systems, cordless phones, and some satellite networks. In the case of the cellular and cordless phones, the path from a mobile user to the public switched network (or vice versa) involves one wireless "hop" (transmission/reception pair).

Cellular telephony requires a fixed supporting infrastructure that includes base stations and land lines that interconnect the base stations with each other as well as to the rest of the PSN (public switched network). For a small satellite terminal, such as the mobile phones used with the geostationary American Mobile Satellite (AMSAT), a mobile user connects to a gateway in two hops—one hop up to the satellite repeater, and a second hop down to the gateway terminal (Earth station). The gateway provides a connection into the PSN. The same applies to the soon-to-be-available low Earth orbit (LEO) satellite systems such as Globalstar and Iridium, except that Iridium has cross-links between satellites and can thus complete mobile-to-mobile calls without using land connections.

From the standpoint of military operations, supporting ground infrastructure matters only when it is in or near the theater of war:

- If not already in place, this infrastructure must be brought in, i.e., some airlift capacity will be required.
- Manpower (skilled technicians) must be available for equipment setup and checkout.
- In-theater infrastructure must be protected against sabotage and accidental damage.

(2) Wireless systems in which users communicate directly through a satellite or satellites. Some military satellite networks (e.g., DSCS) use fairly large satellite terminals. Mobile terminals having sufficient EIRP (Effective Isotropic Radiated Power) and G/T (ratio of antenna gain to effective system noise temperature) and lying within the same satellite antenna footprint can communicate directly to one another through the satellite repeater via two hops. Until relatively recently, communications between small mobile satellite terminals required four hops—two hops to reach a satellite hub, and another two hops (again using the satellite as a repeater) to reach the destination terminal. The new LEO satellite systems being developed will utilize satellites with large antennas and high satellite transponder power levels in order to support direct communications between mobile users.[4]

(3) Wireless data networks that are fully mobile, i.e., any supporting infrastructure is also mobile. No such commercial wireless data networks currently exist. The Army's Tactical Internet is fully mobile. The Mobile Subscriber Equipment (MSE) currently uses a transportable supporting infrastructure, but it may migrate to one that is fully mobile. MSE "mobile subscribers" use the AN/VRC-97 mobile subscriber radio telephone (MSRT), which is a VHF digital radio that supports voice communications (U.S. Army, 1990).

---

[3]"Wired connections" today refer to anything that is not wireless, including the twisted pair wiring in telephone local loops, coaxial cable, and optical fibers.

[4]Access to a gateway is required when a connection is being established in order to verify that users are known to the system and have paid their bills.

CECOM is currently investigating wireless network concepts involving repeaters on UAVs (unmanned aerial vehicles). This concept is described by Sass (1997). A major advantage of the UAV-based relays over satellites is that they can be moved to any location where communications are needed.[5] Furthermore, UAV relays would probably be under the control of the theater commander, whereas satellites are not.

Another concept involves cellular base stations on mobile vehicles; each vehicle would carry an antenna on a tall (e.g., 10-meter high) mast, and would provide connectivity to users in its vicinity. High-capacity microwave trunks could be used to interconnect the mobile base stations, which would be necessary in order to provide connectivity to users served by different base stations. The high-capacity trunks would form a mobile network backbone; mobile-to-mobile connections would involve a single hop into or out of the backbone at each end. The base station vehicles would move with forces, but might have to stop moving in order to operate (see discussion below).

(4) Wireless systems with no supporting infrastructure other than the mobile nodes themselves. Such fully mobile networks are called either *mobile peer-to-peer networks* or *mobile mesh networks*. This technology is an outgrowth of the earlier DARPA-funded packet radio network research of the 1970s and 1980s.

1.2.3 A Two-Dimensional Classification Based on Mobility Factors Only

Consider the general problem of providing connectivity to mobile users through a supporting infrastructure of base stations. One could use a single base station capable of covering the entire area (assuming that this is possible), or a number of base stations, each covering a smaller area. To make a network with multiple base stations behave like a network with only a single base station, one must interconnect the base stations and design the network so that connections are maintained when users move across the boundaries of base station coverage regions ("cells"). The transfer of a user connection from one base station to another is called a *handover*. Furthermore, base stations must track the locations of mobile users even when they are not connected so that connections can be established to them at any time. All of this implies considerable complexity. Untethered mobility with an infrastructure of fixed base stations is best exemplified by the cellular telephone networks.

Various alternatives that achieve lower complexity by sacrificing some functionality are possible. One of these low-complexity alternatives, which we refer to as *tethered mobility*, requires that a mobile node remain within the coverage area of the same base station for the duration of a connection. Cordless telephones represent an extreme form of tethered mobility in which the handset can only be used in the vicinity of a specific base station, i.e., it cannot communicate with other base stations.

---

[5]Geostationary satellites can be moved in order to support a local surge in demand (or to fill a void caused by the failure of another satellite), but this requires a large expenditure of station-keeping fuel and a concomitant reduction in the useful life of the satellite. Furthermore, commercial operators of a satellite on which the DoD has leased transponders would almost certainly not move the satellite to accommodate needs of the DoD users at the expense of other users of that satellite. Thus, although geostationary satellites have sometimes been relocated as an emergency measure, this is not standard practice and should not be counted on as a remedy for capacity shortfalls.

Using a single base station to cover the entire area of interest offers significant advantages in terms of reduced protocol complexity (there is no need for handovers) and reduced computational load. There are, however, several major drawbacks that tend to outweigh these benefits:

- As the size of the area to be covered increases, the required base station antenna height increases (in order to be able to achieve line-of-sight to the mobile users). If the area to be covered is sufficiently large, then it might be necessary to put the base station on a satellite.
- As the size of the area to be covered grows, the EIRP requirements of the base station and of the mobile users increase.
- When a large number of base stations are used, base stations that are not in close proximity can use the same frequency spectrum without interfering with one another. This frequency reuse allows for increased system capacity.
- In a military network, a single base station that covers a large area becomes a critical node, as well as a high-value (and highly visible) asset that is an attractive target for the enemy.

Table 1 presents an alternative two-dimensional classification of mobile wireless systems in which one axis indicates the level of mobility of the supporting network infrastructure (if there is a supporting infrastructure) and the second axis indicates the level of mobility of the users with respect to any infrastructure. Note that there are a total of nine boxes in the interior portion of the table but only eight categories, because one box (marked with an X) corresponds to an impractical combination. One or two examples are provided for each category.

Table 1

A scheme for classifying the mobility aspect of wireless communications networks

	Mobility of user equipments with respect to any in-theater ground infrastructure		
Mobility of the in-theater network ground infrastructure	A Wired connections	B Tethered wireless	C Fully mobile
0-fixed	wired telephone network	cordless telephones	cellular telephones
1- transportable	MSE users with wired connections	MSE users with mobile subscriber radio terminals	cellular system with military version of base station (Ericsson)
2- fully mobile, or no in-theater ground infrastructure	X	network with single UAV relay	packet radio network, LEO satellite systems (Iridium, Globalstar)

Tethered Mobility with Fixed Base Stations

In a tethered mobile communications network, users are constrained to operate within range of a single base station or "access point" for the duration of a connection; the base station is typically fixed. As mentioned before, cordless phones represent an extreme form of tethered mobility. Conventional wireless LANs (WLANs) are another example. WLANs typically operate over very limited ranges, e.g., a single large room or adjacent smaller rooms.

Although tethered mobility might seem unattractive, there are useful military applications, e.g., for voice communications and computer connectivity within and around command posts. Tethered mobility is in general much easier to implement and therefore cheaper than full mobility because a fully mobile network with base stations must be able to perform handovers. Tethered wireless systems is an area with a substantial commercial market, and it is already the subject of intense research and development activity.

Fully Mobile Networks ("Comm on the Move")

These networks do not depend on user proximity to fixed ground infrastructure. Examples include packet radio network networks, which have no fixed ground infrastructure, and satellite communications, which could make use of either military satellite systems or the soon-to-be-available commercial systems such as Iridium, Globalstar, and Teledesic. Another interesting alternative involves networks based on airborne relays.

An interesting intermediate case between transportability and full mobility (full mobility is sometimes called "comm on the move") is where the base stations (or user equipments) are mobile but must stop in order to operate. Cellular base station antennas can be placed on ground vehicles and erected and collapsed using a telescoping mast. In order to provide good coverage, a base station antenna height of at least 10 meters (33 feet) is desirable. However, when the vehicle is moving the mast must be lowered to avoid damaging it. (Some communications equipments cannot operate while moving because of the need to accurately point the antenna.) Thus, there is still some dead time when the base station moves, but much less than for transportable equipment. The dead time might be a significant problem for operations that require continuous availability of communications, unless a given user can communicate through any of several base stations, and at least one of these is available at any given time. For operations against an enemy who has direction-finding equipment, dead time might be less problematic than the need to remain at a fixed location for a significant period of time. Comm on the move permits continuous motion, which in turn reduces the risks of detection and direction finding, which are significant risks for units operating close to the forward line of troops (FLOT) or within range of enemy artillery.

1.3 Circuit-switched Networks and Packet-switched Networks

In this subsection, we briefly review some basic network concepts; readers who understand the terms "circuit switching" and "packet switching" may wish to skip directly to Section 2. There are two basic approaches to the management of resources in communications networks:

- Circuit-switched networks. In circuit-switched networks, a path is established from point A to point B, and a fixed data rate is reserved on each link of that path. The reserved data rate, whether used or not, is dedicated to that connection until the connection is terminated.
- Packet-switched networks. In a packet-switched network, information is broken into segments called *packets* or *cells* that travel through the network in an independent fashion and are eventually reassembled at the destination. In its simplest form, packet-switching does not involve the reservation of resources for any connection. Store-and-forward switching is another name for packet switching.

In the public switched network (PSN), voice connections are handled via circuit-switching, with a dedicated data rate of 64 Kbps (or 32 Kbps) in each direction. The exchange of status and call control information among switches in the PSN is handled via packet switching. Some switched data services (e.g., ISDN) are handled via circuit switching, while others (e.g., frame relay and ATM) are handled via packet switching. The PSN is thus a hybrid circuit/packet-switched network. The Internet, on the other hand, is an entirely packet-switched network.

Circuit switching tends to be inefficient for bursty traffic such as messages and short data transmissions. If one establishes a circuit and then terminates the circuit each time a message is sent, the overhead associated with setup and termination may be much longer than the actual time to transmit the message. If one opens a circuit from A to B, and leaves the circuit open so that messages can be sent immediately whenever they are generated, then the circuit will almost certainly be unused most of the time, resulting in even worse inefficiency.

In the past, one of the major benefits of circuit-switching over packet switching (for wired networks) was consistent and low end-to-end delay once a connection has been established. Because of this, circuit-switched networks seem ideally suited for connection-oriented traffic such as interactive voice and video that has real-time delivery requirements. In traditional packet-switched networks, different packets of a given stream may take different paths through the network. Delay varies from packet to packet because transmission speeds of links can vary, because different paths may involve different numbers of hops, and because queueing delays can vary for successive packets even if they follow the same path.

The delay variability of the traditional packet-switched network is acceptable for such non-real-time traffic as e-mail and file transfers, and even for most Web browsing. Furthermore, because there is no reservation of bandwidth for connections, information transfer tends to be very efficient in traditional packet-switched networks. Such networks are, however, not well suited for traffic with real-time requirements. For connection-oriented traffic with no real-time requirements, such as file transfers, missing packets can be retransmitted, and packets that are received in the wrong order can be re-sequenced. The TCP protocol uses a "sliding window" to keep track of missing packets, and it holds packets that are received out of proper order until missing packets are retransmitted, so that all packets can be delivered in order. For voice, however, the maximum acceptable delay (if the interactive quality is not to be compromised) is about 100 ms, so that one cannot afford to wait for retransmission of missing packets. Furthermore, the packet error rate must be controlled, since packets containing errors must be discarded, and gaps of more than about 50 ms are noticeable to the listener and result in poor intelligibility if frequently occurring.

In recent years, packet-switched network protocols have been developed that support not only non-real-time traffic, but also (like circuit-switched networks) real-time connection-oriented traffic. The best known of these protocols is asynchronous transfer mode (ATM). ATM moves data in fixed-length cells each containing 48 bytes of user data. Although ATM is essentially packet switching at the lowest level, it is able to support a variety of types of traffic, including connection-oriented traffic with real-time requirements, and can provide quality-of-service (QoS) guarantees in wired networks. ATM achieves this by requiring that successive cells in a given connection follow the same path ("virtual circuit") through the network, by reserving resources at switches, by regulating cell flow rates associated with different types of connections, and by admission control (rejecting new connections whose service requirements cannot be satisfied because of the existing network load).

Current Internet protocols provide no guarantees other than TCP's guarantee of eventual delivery under reasonable conditions. However, a protocol known as the ReSerVation Protocol (RSVP) is under development that will enable support for predictive or guaranteed QoS when it becomes available (Corson and Macker, 1996).

> Messages and short data transmissions are a large and growing fraction of tactical network traffic. Because circuit switching is inefficient for such traffic, and because packet switching is also more flexible than circuit switching, there is a growing consensus that future military networks will be based entirely on packet switching.

Documents such as the Joint Technical Architecture (DoD, 1997) show that the Army is committed to the integration of standard protocols and commercial technology such as TCP/IP and ATM (see also Sass, 1997, and U.S. Army Signal School, 1997). The Army should be commended for this resolve, and for the steps already taken to implement it. There is, however, a tension between these goals and the desire to retain legacy systems. When commercial protocols are used either with legacy military systems that were not designed to support those protocols or in environments very different from those for which the protocols were designed, performance problems that are difficult to predict and harder to remedy are likely to occur, as was seen in recent Tactical Internet experiments.

> Mismatches between Army requirements and commercial standards/practices will have to be addressed by changing the requirements, departing from commercial practice, inducing (to the extent possible) necessary changes to and extensions of commercial standards, or (most likely) a combination of all of these.

2. Performance Measures and Design Tradeoffs

In this chapter we discuss some of the more important performance measures and tradeoffs involved in the design of communications networks, and we explain why commercial and military system designers have tended to make these tradeoffs differently.

2.1 Introduction

The performance of a communications system (or any other system) depends on *design parameters* whose values can be selected by the system designer and *environmental parameters* over which the designer has no control. The relationship between these parameters and performance metrics of interest is usually complex. In general, changing any single design parameter tends to impact all performance metrics of interest, and simultaneously changing multiple design parameters typically affects performance metrics in ways that cannot be predicted from knowledge of the single parameter effects alone.

The goal of the design process is select the design parameters so as to achieve specific performance levels (or the best performance possible) subject to constraints on system cost (cost can thus be viewed as another performance metric). Some of the choices the designer must make are essentially discrete or integer valued, i.e., a selection among a small (or at least finite) set of alternatives. The three-way divide between narrowband, direct sequence spread spectrum, and frequency hop spread spectrum is an example of a situation where such a choice must be made. Other design parameters are essentially real valued. For example, antenna size and transmitter output power can take on values from a continuum.[6]

The wireless system design problem is difficult for several reasons:

- The designer is faced with a huge design space (each design parameter can be thought of as one dimension in a multidimensional space). Exhaustive exploration of this space is typically impractical. Thus, the designer must rule out many alternatives early in the design process on the basis of experience (his own or others') in order to consider a smaller, more manageable set of alternatives that can be evaluated through simulation.
- Current simulation tools tend to (at best) accurately model either the ISO physical layer (layer 1) on a single-link basis, or the middle ISO layers (2–4) for networks involving multiple nodes, but not both at the same time.
- Even without detailed modeling of the physical layer, high-fidelity simulations of large networks tend to require large amounts of computation. One cannot scale down networks for purposes of performance evaluation because the behavior of networks involving small numbers of nodes may be very different.

---

[6]Of course, unless one is willing to use custom components and subsystems, manufacturer's catalogs for such items as antennas and high-power amplifiers typically offer only a discrete set of choices. Thus, the designer is in practice limited, at least in the selection of components and subsystems, to a finite but very large set of alternatives.

- The external environment in which a system must operate is often highly uncertain. Terrain type, presence of interfering equipments, jamming, and other external factors can all impact performance, but are difficult to accurately characterize and model (see Chapter 5). In the case of jamming, uncertainty about the threat is a major issue.

Military and commercial communications systems designers tend to take different approaches and reach different results primarily because (1) the expected operating environments are different, (2) the business practices and economics (including economies of scale) are different, and (3) certain performance attributes, e.g., robustness against jamming and low probability of detection, are of concern only for the military.

2.2 Criteria for Comparing Mobile Wireless Systems

Performance requirements of communications networks depend on a variety of factors, including the types and quantities of traffic to be carried, the required availability and responsiveness of the system, the operating environment, and acceptable costs for the infrastructure and user equipment segments of the network. Some performance measures are specific to certain types of networks, or to certain types of traffic, and make no sense in other contexts.

2.2.1 High-level Performance Measures

We discuss "high-level" performance measures first, because these are the measures the user is most directly aware of. Performance measures specific to circuit-switched networks and to connection-oriented traffic on some packet-switched networks (e.g., ATM networks) include:

- blocking probability. This is the probability that a request for a circuit or connection fails because the system cannot accommodate additional circuits/connections.[7] Note that in the current Internet, a connection can fail because a host is unreachable, but it cannot be blocked because of excessive congestion (it can, however, time out).
- circuit/connection setup time. This is the time to set up a circuit or connection when blocking does not occur.

For connectionless traffic on packet-switched networks, the performance measures of interest are:

- probability of (successful) delivery.
- end-to-end delay (for packets that are delivered). For comparison purposes, the delay distribution is often reduced to a single statistic (e.g., the mean or 95th percentile).

For any networks where data is transmitted without retransmissions, error rates are critically important. Error rates are typically low for wired connections, but vary enormously for wireless links. Depending on the formatting and content of the data, the relevant measure would be

---

[7]In the PSN, blocking at the local office is indicated by absence of dial tone.

- end-to-end bit error rate (BER), symbol error rate, packet error rate, message error rate, or line error rate (for video).

For connection-oriented traffic with real-time requirements (on packet-switched networks):

- the packet delay and the delay *jitter* (variation in delay from one packet or cell to the next).

For connection-oriented traffic without real-time requirements, the user is probably most sensitive to the:

- connection throughput or total delivery time.

Note that these two quantities are related in a simple fashion when connection throughput is constant, since time to send a message or file equals the length in bits divided by the connection throughput in bits per second (plus the connection setup time).

2.2.2 Low-level (Physical Layer) Performance Measures

The high-level performance measures presented in the previous section depend on the network protocols, the quantities of traffic, the operating environment, and (of course) the performance of the underlying hardware. We now discuss some of the more important low-level, or physical layer,[8] performance measures.

- Average link throughput (user data bits per second). If forward error control coding is being used, the user bit rate will be lower than the channel rate (the ratio is the code rate). If error detection is used, the link throughput is further reduced by the multiplicative factor of the packet loss rate (packet loss probability). Suppose, for example, that the channel bit rate is 19,200 bps, a rate-3/4 forward error correction code is used, and the packet loss rate is 10 percent. The average link throughput is then 19,200 bps $\times$ 0.75 $\times$ 0.9 = 12,960 bps.[9]
- Average terminal power consumption (Watts) and antenna size. Average power consumption is important because it determines the operating time per charge for a given battery type and weight.
- Link error rates: bit error rate (BER), symbol error rate, and packet error rate. The maximum acceptable BER depends on what type of information is being transmitted over the link. For some types of vocoded voice (e.g., 16 Kbps CVSD), BERs as high as 0.01 might be acceptable.[10] For compressed imagery, link error rates of 10^{-6} or even lower might be required (a single bit error could cause the loss of the entire image).
- Maximum user density (number of active users per km^2).
- Spectral efficiency (bits per second per Hertz).

---

[8]Note that the term "link" as used by hardware engineers typically refers to OSI layer 1 (the "physical layer"), rather than OSI layer 2 (the "link layer").

[9]This calculation assumes either that the error detection code rate is approximately unity or that this code rate is lumped together with the error correction code rate.

[10]Depending on the level of acoustic background noise competing with the speaker's voice, and other factors.

Some of the performance metrics in the above list appear to duplicate high-level performance metrics. However, link throughput and error rate are not the same as end-to-end throughput and error rate (e.g., in a system without error control, the error rate of each link of a multilink path will contribute to the end-to-end error rate).

Users, system designers, and long-range planners tend to rank the above performance measures differently. The individual user cares about end-to-end error rate, since this directly relates to connection quality. Whether this error rate is associated with a single link (one-hop transmission) or the compound effect of errors on several links (multihop transmission) is of concern to system designers, but of no real interest to users.

2.2.3 Military-Unique System Requirements

The military has some unique requirements that tend to drive the design of military communications systems toward solutions that are markedly different from commercial systems. The most important of these differences involve: (1) low probability of detection, (2) resistance to jamming, (3) precedence and perishability, (4) electromagnetic compatibility, (5) interoperability with legacy systems, and (6) security.

2.2.3.1 LPD and Resistance to Jamming (A/J)

Low probability of detection (LPD) is critical for activities such as reconnaissance because it reduces the risks to forward spotters, and it is important in any situation where direction-finding equipment might be employed to advantage by the enemy. Measures that both increase resistance to jamming (A/J) and reduce the probability of detection include:

1. Use of direct sequence spread spectrum (DS SS), frequency-hop spread spectrum (FH SS), or a combination of the two with secure spreading sequences (pseudo-random spreading sequences generated from a secret key).[11] Note that commercial spread spectrum systems use nonsecure spreading sequences and thus have no A/J or LPD advantages over nonspread systems. Also, even if the spreading sequences were secure, the spread bandwidths and spreading gains are simply too small. For example, the IS-95 standard for cellular telephony uses a spread bandwidth of approximately 1.5 MHz. For a 900 MHz carrier frequency, this corresponds to a bandwidth/carrier ratio of less than 2 percent, which can easily be covered by a conventional jammer using a single ordinary klystron tube. At 900 MHz, a spread bandwidth of 150 or 200 MHz would provide some benefit against such a jammer. See Appendix B, Section B.2, for a brief overview of spread spectrum techniques.
2. Use of directive antennas with narrow main beams and low sidelobe levels. Using a directive transmit antenna reduces the probability of detection by unintended receivers (directive transmit antennas also reduce the "friendly" interference power seen by other receivers). Using a directive receive antenna makes jamming more difficult.[12]

---

[11]To be secure, the pseudo-random ("pseudo-noise") sequence must be of sufficient length to not repeat before the next rekeying. The key space must be large enough to withstand any real-time combinatorial attack (note that the requirements for encryption are much more stringent because there is potential value to an attacker even if he cannot decrypt in real time).

[12]For parabolic reflector antennas, an increase in the amount of edge taper on the illumination pattern can substantially reduce near sidelobe levels (e.g., to 40 dB below the peak gain, as compared with 20 dB below

3. Use of relays, e.g., a relay on a UAV. A UAV relay permits ground terminals to transmit at lower power levels (assuming that the range to the relay is less than the range to the intended receiver) and to direct their transmissions at high elevation angles; both of these measures would tend to impede detection and direction finding by enemy ground platforms.

Measures that are beneficial against jamming but of no benefit (or possibly detrimental) for LPD include:

4. Operating with increased average transmitter power. This increases the probability of detection.

5. Use of strong FEC coding such as concatenated block and convolutional codes with an interleaver to break up error bursts caused by pulsed jamming.

6. Use of modulations that are less vulnerable to jamming. For single-carrier applications, maximal spectral efficiency can be achieved by using high-order nonorthogonal modulations such as 1024-QAM (quadrature amplitude modulation). However, such modulations are highly vulnerable to jamming. Binary GMSK (Gaussian minimum shift keying), although less spectrally efficient than 1024-QAM, is much less vulnerable to jamming while achieving reasonable spectral efficiency.

7. Use of advanced signal processing techniques, e.g., generation of jammer side information (an estimate of the likelihood that any given demodulated symbol was affected by jamming). This information can be used to aid the FEC decoder.

8. Use of adaptive antennas. Adaptive nulling can be extremely powerful against jammers that are well separated in angle from the desired signal source(s). However, adaptive nullers that can cope with multiple jammers entail high complexity and high cost. For a review of adaptive antenna arrays, see Godara (1997).

One other technique that is worth mentioning increases LPD, but typically provides no A/J benefit:

9. Limiting the durations of transmissions.

Some commercial wireless systems are now using spread spectrum techniques; examples include the IS-95 digital cellular CDMA standard, and two of the three waveforms in the recently approved IEEE 802.11 standard for wireless LANs. However, spreading gains[13] tend to be much smaller in these commercial waveforms than the spreading gains that would be used in a military system for protection against broader-band jamming. Even more problematic, none of the commercial systems uses secure spreading sequences because of the key distribution problem.

---

the peak gain for a conventional antenna). This requires a different feed and also entails some broadening of the main beam and a 1 or 2 dB reduction in the peak gain.

[13]Spreading gain is defined as the ratio of the spread and unspread bandwidths. After despreading, the desired signal becomes narrow band, while the jamming signal is spread over the entire spread bandwidth. Filtering can then be used to remove most of the jammer power.

Data communications, because of their comparative brevity, tend to be more conducive to LPD than voice communications. Furthermore, because data, unlike interactive voice, is amenable to ARQ (error control via re-transmissions) and variable-rate forward error control,[14] data transmissions can be made far more robust against jamming than voice transmissions.

2.2.3.2 A/J Benefits of Two Types of Spread Spectrum

Although one cannot make completely general statements about the A/J benefits of competing combinations of waveforms and signal processing, one can compare specific waveform/signal processing combinations against specific types of jamming. For the sake of simplicity and brevity, we consider only two types of spread spectrum—slow frequency hopping (SFH) and direct sequence (DS)—and two generic jamming cases:[15,16]

- high-power pulsed noise jammers that operate over the entire frequency band of the desired signal (these may be called either "pulsed full-band" or "partial-time full-band" jammers), and
- lower-power noise jammers that operate continuously, i.e., at 100 percent duty cycle, but cover only part of the desired signal frequency band ("continuous partial-band jammers").

For both SFH and DS systems (see Appendix B for a brief overview of spread spectrum techniques), full-band pulsed jamming tends to produce bursts of errors at the demodulator output (one burst for each jammer noise pulse). In the case of continuous partial-band jamming, however, the jamming has different effects on the two systems, producing "random" (i.e., nonbursty) errors for the DS system,[17] but short error bursts for the SFH system.[18] Many simple error control coding schemes (e.g., convolutional coding only) work better with random errors than bursty errors (see Lin and Costello, 1983, and Blahut, 1983). However, concatenated coding schemes have been developed that are highly efficient at correcting bursty errors; see Frank and Pursley

---

[14]Retransmission delay is incompatible with interactive voice. Variable-rate error control coding, which adapts the code rate in response to changing channel conditions, can be used with interactive voice only if a special variable-rate vocoder is used.

[15]We do not consider continuous full-band jamming because: (1) Against spread-spectrum signals that are spread over a large enough bandwidth to have significant A/J, the average EIRP required for such jamming tends to be very high. As a consequence, these jammers require large antennas, as well as diesel fuel generators. Such jammers (which begin to look like strategic jammers) are not mobile and thus can easily be located and destroyed in a tactical setting. (2) If the jammer can jam the entire band all the time, and do so with sufficient power, then no error control scheme will avail. Consequently, this case is not of interest

[16]Other types of jamming that are important include fixed tone jamming, swept tone jamming, and follower jamming. Follower jammers attempt to quickly determine the carrier frequency of a frequency hop system at the start of each hop and then produce either a tone or narrowband noise in the vicinity of that carrier. For operation against a follower jammer, the hop dwell time of a frequency hop system must be shorter than the relevant signal propagation times and measurement delays.

[17]The direct-sequence despreading processes, which precedes the demodulation, converts narrowband and partial-band noise into full-band noise, so that all symbols are corrupted to some degree by the jamming power.

[18]When a hop falls in the jammed part of the band, there is a burst of errors at the demodulator output; the length of this burst corresponds to the number of channel symbols per hop. If the jammed fraction of the band is small, then consecutive jammed hops will be rare.

(1996) for a concatenated coding scheme that is optimized for slow frequency hopping in partial-band interference.

Because one can always use interleaving with a sufficiently long span to convert bursty errors into random errors, it might appear that the difference between bursty errors and random errors is unimportant, but this is not the case. With pulsed jamming against either SFH or DS, and with continuous partial-band jamming against SFH only, jammer energy is concentrated on selected symbols. With continuous partial-band jamming against DS, jammer energy[19] is divided among all symbols. When jamming energy affects only a fraction of the symbols, a suitably designed receiver can (with high probability) determine whether the jamming has affected any given symbol.[20] This symbol-by-symbol information about the presence or absence of jamming is known as *jammer side information* or *jammer status information* (JSI) and can be generated in any of several ways. For further discussion of JSI, see Pursley (1993), Baum (1992), and Pursley and Wilkins (1997). As these papers show, a suitably designed error control decoder with access to JSI can substantially outperform a conventional error control decoder when the jammer energy affects symbols unequally.

Since low-duty-cycle pulsed jamming is not a serious threat for any system (spread or unspread) with strong coding, continuous jamming is arguably the more important threat. For operation against a continuous partial-band jammer, SFH appears to be superior to DS (for the DS system, JSI is of value only against pulsed jammers).[21] Given uncertainty as to which type of jamming one may face, it seems fair to say that (in combination with suitable receiver signal processing), slow frequency hopping is a more effective A/J technique than direct sequence. An additional benefit of SFH is the ability to spread over wider bandwidths than is possible with DS (see, e.g., Torrieri, 1997). For supporting information, as well as other comparisons of SFH, DS, and other spread spectrum schemes, we refer the reader to Fiebig (1998), Torrieri (1997), and Gass and Pursley (1997).

2.2.3.3 Precedence and Perishability

Military networks must be able to offer different grades of service to traffic on the basis of *precedence* level (priority), which indicates importance, and *perishability*, which indicates when the information must be received in order to be of value. Optimal handling of precedence and perishability information is especially important when a network becomes congested. In packet-

---

[19]We refer here to only that portion of the jammer power that falls into the comparatively narrow bandwidth of the despread signal; the rest is removed by filtering.

[20]A jammer "affects" a given symbol if it changes the unquantized demodulator inputs associated with that symbol. The demodulator may make a correct decision on an affected symbol, but is more likely to make a wrong decision than for an unaffected symbol.

[21]This is a bit of an oversimplification. Performance of a SFH system may be sensitive to the band fraction jammed. If the jammer has either (1) good information about the target waveform and about the geometry or (2) some way of assessing its effectiveness in real time (e.g., by monitoring the downlink corresponding to a jammed bent-pipe satellite uplink), it can adjust its bandwidth to cause the maximum harm. For the DS system, on the other hand, the band fraction that is jammed (subject to fixed total jammer power) has virtually no impact on error rates, leaving the jammer with no scope for game playing. Thus, in situations where the jammer is maximally effective against the FH system when jamming a small fraction of the band, i.e., against a system with weak coding or no coding, DS spreading may give better performance than SFH. (This footnote is based on a private communication from Dr. Donald Olsen of the Aerospace Corporation.)

switched networks such as the Internet, network status and control information and certain types of traffic receive expedited service, but there are no special classes of privileged users.

The commercial world has tended, for the most part, to reject the idea of different grades of service for different customers. Commercial packet-switched network protocols, including the TCP/IP protocol suite, ATM, and frame relay, do not currently provide any mechanism for special handling on the basis of precedence and perishability.[22] In commercial circuit-switched networks such as the PSN, there is no automatic mechanism for terminating low-precedence calls in order to free circuits for higher-precedence calls. Such capabilities are essential for tactical networks.

2.2.3.4 Guaranteed QoS Versus High Probability of Timely, Intelligible Communications

Guaranteed QoS makes sense for wired networks with stable topologies and constant link capacities but is almost certainly unrealistic for fully mobile wireless networks, even without the added factor of hostile enemy actions such as destruction of nodes and jamming of links. Furthermore, guaranteed QoS requires admission control, which is unacceptable on the battlefield except as a last resort. That is:

Tactical users need immediate access (for voice) or high probability of message delivery (for data) at the best quality available under current conditions; the commercial world is moving in the opposite direction—in ATM networks, quality of service is to be maintained by limiting the numbers of users who can simultaneously communicate.

For robust tactical wireless voice communications, one needs a radio in which the vocoder responds to feedback from the link level about the data rate currently available, increasing or decreasing its output rate accordingly. Lower vocoder output rates permit lower forward error control (FEC) coding rates; with lower-rate FEC coding, the receiver can correct a higher percentage of errors. At the lower vocoder output rates, the reconstructed speech might begin to sound somewhat mechanical but should still be intelligible. Under extremely stressed conditions, the radio might switch from the vocoder to a speech recognition chip, convert the speech to text, and transmit it as a text file.

The point is that in critical situations, the ability to get something intelligible through in a timely fashion is probably the most important tactical user comm requirement.

2.2.3.5 Electromagnetic Compatibility

For some military platforms, for command posts, and vehicles moving in formation, electromagnetic compatibility (EMC) can be a problem because of interference between multiple equipments operating in close proximity (possibly on the same platform); this is the so-called "co-site interference problem." These equipments might be radios or radars, or equipments for naviga-

[22]The admission control part of ATM is implemented in software and could therefore in principle be modified to handle precedence in any of various ways, including the dropping of existing connections. But implementing such a change would require a move from unconditional QoS guarantees to something more flexible.

tion, IFF (identification friend-or-foe), or electronic support measures. EMC can often be achieved by selecting operating frequency bands that do not overlap (although harmonics can in some cases still cause interference). For equipments operating on the same platform, the best approach sometimes involves the use of blanking intervals, e.g., the input to a radio receiver front end might be briefly blocked during the short interval when a nearby radar is transmitting. Commercial communications equipments that don't provide the flexibility for adjustment of frequencies or for blanking may be unacceptable for military applications.

EMC issues are also part of the reason for the military preference for FH SS over DS SS. With DS SS, there is essentially no control over the transmitted power spectral density.[23] With FH SS (or combined FH SS/DS SS), on the other hand, the *hop set* (set of allowed frequencies) can be chosen so as to avoid certain frequencies. JTIDS, for example, avoids frequencies that are used for navigation and IFF. Some frequency hopping radios (SINCGARS is an example) allow for programmable control of the hop set.

2.2.3.6 Security

Security is often cited as an additional military-unique performance requirement. This is not entirely accurate, however, because the business world is becoming increasingly concerned about the protection of information, and widespread commercial use of strong encryption and authentication (digital signatures) seems inevitable. Version 6 of IP will include support for both encryption and authentication. Still, there are several security requirements that appear to be unique to the military. These include:

- Even if all user data is encrypted, TRANSEC is needed to protect signaling and control traffic; without this, an eavesdropper could do traffic analysis. The commercial world never worries about this, even if they do protect the data by encryption (COMSEC).
- Secure multicast with frequent changes of multicast group membership may require more complex mechanisms for key generation, distribution, and authentication.[24]
- Military wireless networks must be capable of surviving the capture of radios by the enemy.

2.3 Capacity and Spectral Efficiency Issues

2.3.1 Not All Capacity Is Created Equal

A deceptively simple question in mobile communications networking is this: "How does one assess the 'capacity'[25] (maximum throughput) of the network?" There are a number of different ways of doing this, and the results can vary significantly, depending on:

---

[23]Filters could be used to shape the transmitted power spectral density, but the filters would add weight and complexity. The filtering might also adversely affect the performance of the system.

[24]The commercial world also needs to solve the problem of authenticating users who join a multicast, since otherwise it would be impossible to have private conferences, pay-per-view events, and the like. However, the military authentication requirements are more complex.

[25]Our use of the term "capacity" here is consistent with the popular usage of the term, but not with the formal information theoretic definition. See, for example, McEliece (1977).

1. whether one considers per-user throughput or total network throughput. Total network throughput is (at most) the sum of all user transmit data rates at a given time.
2. the maximum tolerable bit error rate (BER) or message error rate (MER). For systems in which users or nets generate mutual interference, the maximum number of simultaneous transmissions increases with the maximum error rate that one is willing to accept. Thus, total network throughput also depends on the maximum error rate.
3. the geographical dispersion of the transmitters and receivers. Again, in networks where mutual interference can occur, one might expect that some geometries will be more favorable and others less. For cellular and related types of networks, it may be more appropriate to think in terms of the maximum average number of transmitters per square kilometer (or throughput per square kilometer), rather than per-user throughput or total network throughput.

As an example consider the chart below, taken from the CMA Communications Mix Study. The chart indicates that the JTIDS link (i.e., per user) data rate can range from 28.8 Kbps to 238 Kbps.

Note that the chart is misleading in two ways:

(1) A single JTIDS net is shared in a TDMA fashion by all users on that net. The JTIDS architecture is a full-mesh (all-to-all) communications network, i.e., any transmission is received directly ("single hop" transmission) by all receivers who are listening on the same net. Thus, the maximum throughput of 238 Kbps can be achieved by a single user only if no one else is permitted to transmit on that net.

Assessment of Baseline: Tactical Radios

- **Baseline**
  - » **Many digital and analog radios**
- **Significant capability shortfalls**
  - » **50X data rate increase needed**
  - » **Not all forces are digitized**
- **Inadequate interoperability**
  - » **Many different waveforms**
- **Inadequate Joint real-time sensor-to-shooter capability**
  - » **Translating gateways provide limited, cumbersome solutions**
  - » **JTIDS/VMF provides solution for selected cases but is too costly for wide-scale use**

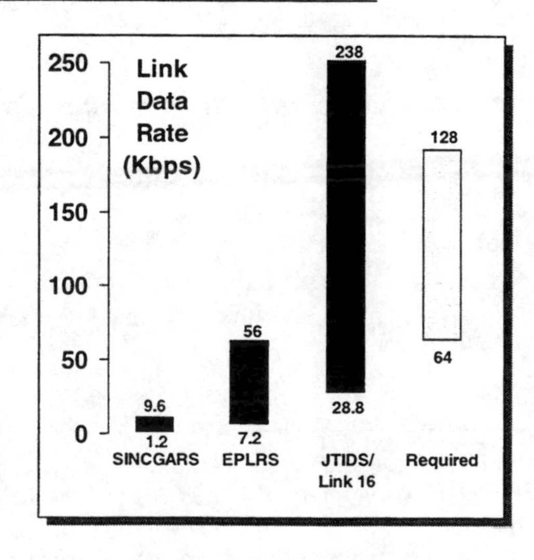

The design and sizing of a network should reflect the type of traffic that will flow over the network. If most of the traffic is broadcast in nature, i.e., it is necessary or desirable that transmissions be received by all users, then a full-mesh architecture like JTIDS can be reasonably efficient. However, for most tactical communications, a large fraction of the messages or voice calls are among small groups of nearby users, where the typical separation distance is small compared to the extent of the entire network. Under such conditions, architectures that make use of shared radio nets,[26] point-to-point links, or a combination of the two, with some communications requiring more than one hop, can make much more efficient use of spectrum and power, as well as offer much higher real capacities.

(2) JTIDS achieves the 238 Kbps data rate only when FEC coding is turned off. However, JTIDS error rates are typically unacceptable without error control. With FEC, the maximum possible throughput for a single JTIDS net is either 115.2 Kbps ("single pulse mode") or 57.6 Kbps ("double pulse mode," which is more robust).

Note that (2) illustrates the important concept that capacity is meaningless by itself. One must consider both capacity and error rate[27] when evaluating whether a given system is adequate for a particular purpose. A 1 Mbps link with an error rate of 10^{-8} may be preferable to a 10 Mbps link with an error rate of 10^{-5} because highly compressed images cannot be transmitted over the noisier link. (In order to send images over the noisier link, one might be forced to either send uncompressed images, which could more offset the higher speed of the faster link, or perform additional error control coding, which might not be feasible.)

2.3.2 Spectral Efficiency

There are growing demands from military users for access to communications. There is an increasing number of users who desire access, as well as a demand for more bandwidth (higher data rates) per user. At the same time, there is increasing pressure on the military to give up frequency spectrum for sale to commercial service providers. Clearly, these forces are in conflict. Assuming, however, that encroachment on military spectrum can be halted, higher capacities can be achieved by a combination of (1) more efficient use of the frequency spectrum and (2) greater exploitation of frequencies at X-band (8–12 GHz) and above. Since even at the higher frequencies spectrum is limited, more efficient spectral utilization will be a high priority for all military systems.

Spectral efficiencies of military communications systems have tended to be relatively poor; efficiencies below 0.1 bps/Hertz are typical, and efficiencies below 0.01 bps/Hertz are not uncommon.[28] Compare this with spectral efficiencies (not counting cell-to-cell frequency reuse) on the

---

[26]Communications in a radio net involves short-range broadcasts to all participants of the net (single-hop communications).

[27]The information theoretic definition of capacity is the maximum information rate, and also equals the maximum error-free user data rate that could be achieved with ideal forward error control (FEC) coding. We are using the term "capacity" as a substitute for "maximum user data rate" (a technically incorrect but widespread usage).

[28]JTIDS has a nominal spectral efficiency of 0.02 bps/Hertz when 30 nets operate simultaneously, but calculations done by this author indicate that operation of more than about 5 simultaneous nets will result in excessive message error rates. With 5 nets operating simultaneously, the maximum spectral efficiency of JTIDS is roughly 0.003 bps/Hertz.

order of 0.1 to 0.5 for commercial cellular systems, with efficiencies toward the upper end of that range for the newer digital cellular systems. Part of the explanation for this is the military need for waveforms with A/J (anti-jam) and LPD characteristics, which are easiest to achieve by sacrificing spectral efficiency. However, this does not explain the spectacularly low spectral efficiency of many military systems. Spectral efficiency is irrelevant from the perspective of the individual system user, and it is of concern to the military system designer only to the extent of supporting the required number of users within the allocated/available spectrum. If the allocated spectrum permits a spectrally wasteful design that allows for minor economies in terms of reduced power consumption and system complexity, the wireless system designer might be tempted to select a spectrally wasteful design. For the long-range planner, who views the frequency spectrum as a scarce resource that must be carefully parceled out to competing systems and groups of users, and who must also worry about the as-yet-undefined requirements of future systems, spectrum efficiency is a much higher priority than it is for the military system designer.

The commercial wireless hardware designer must worry about spectral efficiency because this may ultimately determine whether a system is profitable or not. Spectral efficiency determines the capacity limit of the system, which in turn determines the maximum user base that the system can support, which determines the amount of hardware that can be sold to service providers and users. Or, from another perspective, if system A offers higher capacity than system B and hardware costs are comparable, then system A should be expected to win out over system B in the marketplace. Designers of military systems regard spectral efficiency as less important because market forces, to the extent that they are present, involve primarily the initial acquisition costs of competing systems, and not the life-cycle costs or effective per-user costs, since military users do not pay for communications hardware and services.

Although spectral efficiency can be measured in various ways, bits per second per Hertz is probably the most common. Techniques for increasing spectral efficiency include the following:

1. One can use modulation formats such as binary Gaussian minimum shift keying that are spectrally compact. With spectrally compact modulations, channels in an FDMA system can be packed more closely together because there is less power spillover into adjacent channels. Quadrature phase shift keying, a modulation widely used in military communications systems, is spectrally inefficient.

2. It may be possible to use power control to balance power, so that amplitudes of signals arriving at a given receiver are matched as closely as possible.[29] Power balancing also permits closer packing of channels. For some types of wireless systems, power balancing is impractical. For example, in some military satellite systems, user terminal EIRPs differ by 20 to 30 dB, rain uplink losses vary from 0 to 20 dB, and the satellite receive antenna gain varies by 8 dB from the peak of a beam to the "triple-point" between three adjacent beams. Taking the worst case combination of these, we see that variations of up to 58 dB between different channels are possible; such variations are far too large to be corrected out by power balancing. For systems designed to cope with jamming, it may be desirable that each terminal operate at its maximum output power; this clearly precludes power balancing.

---

[29]In systems with channels of unequal widths, one must balance the spectral power densities (watts per Hertz).

3. Many FDMA systems use fixed-width channels that are sized to support the highest user data rate. For example, MILSTAR MDR uses 10 MHz-wide channels for data rates ranging from 19.2 Kbps to 2 Mbps; this bandwidth is reasonable for the 2 Mbps signal (spectral efficiency of 0.2), but grossly inefficient at 19.2 Kbps (less than 0.002). Dynamic adjustment of channel width, which can be accomplished using digital filtering, permits one to maintain roughly constant per-channel spectral efficiency.

4. In satellite systems and in other networks with repeaters, dedicated assignment of channels to specific users (for broadcast) or pairs of users tends to lead to underutilization. Demand assignment multiple access (DAMA) techniques can be used to dynamically assign channels as needed, permitting a population of users to share the resources. See Feldman (1996) for a survey and comparison of DAMA techniques.

5. Even with DAMA, it is common practice to fence off pools of channels, with each pool dedicated to a specific group of users. This tends to lead to situations where one pool is overloaded while another is largely unused, since demand cannot be shifted between them. Merging pools of channels together avoids this problem.

6. In satellite and other repeater-based systems that use TDMA, a portion of each time slot must be reserved for guard time. This part of the slot is left empty in order to account for uncorrected time delay differences between transmissions arriving from different sources, so that a transmission in slot k will not overlap and interfere with a transmission in slot $k+1$. With accurate knowledge of position and time, transmit times can be adjusted to correct for range differences to the repeater, allowing for guard times to be virtually eliminated.

7. In systems with multiple repeaters (base stations), base stations that are distant from one another can use the same frequencies (or codes, for code-division multiple access systems). Such frequency reuse allows for potentially large increases in system capacity.

The military is impeded from transitioning to more efficient use of the frequency spectrum by the problem of legacy systems. In particular, radios and satellite terminals that use older modulation formats cannot interoperate with ones that use newer modulations. Worse than this, it may not be possible to use these equipments in the same theater of operations because of mutual interference.

2.4 Key Design Tradeoffs

2.4.1 Power Efficiency, Spectral Efficiency, Complexity, and Resistance to Jamming

Consider an ideal band-limited channel of bandwidth W Hertz in which the signal is corrupted only by additive white Gaussian noise (AWGN) having a one-sided power spectral density level of N_0 watts/Hertz. (This simple model does not apply to channels that corrupt signals in more complex ways, e.g., by multipath or mutual interference.) Let E_b denote the energy per information bit at the receiver and let C denote the *channel capacity*, i.e., the maximum average rate at which information can be transferred over this channel. The following formula, which can be derived from Claude Shannon's capacity formula (see Appendix D), relates the maximum achievable spectral efficiency C/W to the signal-to-noise ratio (SNR) E_b/N_0:

$$\frac{E_b}{N_0} = \frac{W}{C}\left[2^{C/W} - 1\right].$$

A graph of this equation, which appears in Figure 1, shows that there is a tradeoff between power efficiency and spectral efficiency. As one can see from the graph, waveforms that achieve very high spectral efficiency (e.g., high-order quadrature-amplitude modulation) require high SNR, while the most power-efficient waveforms (e.g., orthogonal frequency-shift keying) are wasteful of spectrum.

Using the above equation, one can readily find that in order to achieve a spectral efficiency of 6 bps/Hertz, the minimum required signal-to-noise ratio E_b/N_0 = 10.5 = 10.2 dB. Note that although this value of SNR is a practical operating point for many wireless systems, most existing wireless systems that operate at SNRs in this neighborhood actually achieve spectral efficiencies less than 1.0, and many military wireless systems and satellite systems have spectral efficiencies less than 0.1. Thus, there is clearly substantial room for improvement.

The simplest way to achieve higher spectral efficiency is by increasing E_b/N_0, which in turn implies increasing the transmitted signal EIRP in the direction of the receiver (recall that EIRP is the product of transmitted power and antenna gain), increasing the receiving antenna gain, decreasing the receiving system noise figure, or some combination of these measures. The price of simultaneous power and spectral efficiency is a substantial increase in complexity. Nevertheless, combined modulation/coding techniques that achieve fairly high power and spectral efficiency simultaneously have been developed in recent years, and commercial ASICs that implement some of these techniques are now available.

Adding requirements for jam resistance requires some sacrifice of power efficiency, and some additional complexity as well, e.g., to implement error control decoders that use jammer side information (see Section 2.2.3.2). However, as device counts of VLSI chips and MIPS ratings of

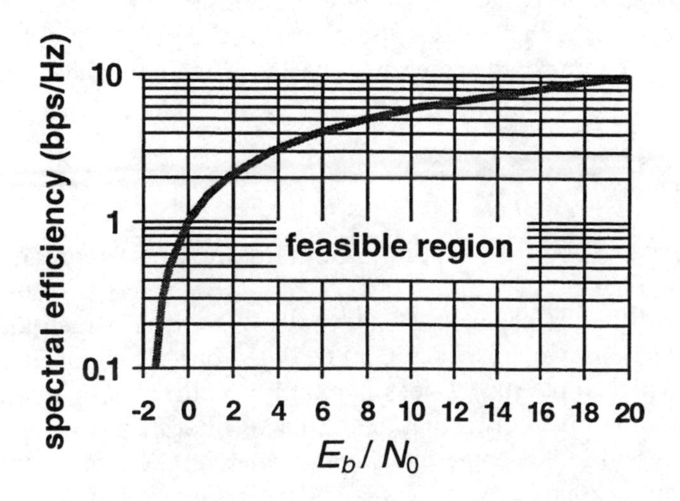

Figure 1—The theoretical limit for spectral efficiency C/W as a function of the signal-to-noise ratio (SNR) E_b/N_0

microprocessors continue to increase, higher levels of complexity are becoming increasingly practical.

2.4.2 Business Practices

The military tends to retain computer and communications equipment in the inventory for relatively long periods. In the commercial world, users are expected to upgrade or replace equipment every 4–10 years. Because volumes of military systems tend to be much smaller than those of commercial systems, the costs of R&D, software, and testing have a much greater impact on the final per-unit cost of military systems than they do for commercial products. These higher costs in turn force the military to try to retain the systems for as long as possible.

2.5 Interoperability Issues

Differences in frequencies of operation, waveforms (modulation and error control coding), protocols, and message catalogs prevent the different radios from interoperating. Interoperability problems have drawn considerable attention in recent years, but current efforts to improve interoperability are largely separate-service efforts that do not address the need for communications interoperability between services and with our allies for joint and combined operations.

2.5.1 Encryption and ECCM

Because cryptographic and ECCM algorithms in tactical radios have been implemented in hardware, and since the relevant devices and specifications are generally not made available to our allies, encrypted and jam-resistant tactical communications are problematic for combined operations.[30] Part of the solution for this problem may be software-based encryption; this permits one to change not only the keys, but also the underlying algorithms. Software-based encryption is practical except perhaps at the highest data rates. Secrecy with respect to the encryption algorithms used in tactical radios is almost certainly a mistake. If we trust the algorithms, then there can be no harm in making them public. If we do not trust them, then scrutiny by academic researchers is one of the best ways to find flaws. This same basic reasoning about the secrecy of encryption algorithms can be applied to the pseudorandom sequence generators used in frequency hopping and direct sequence spread spectrum, but it does not apply to ECCM algorithms in general.

2.5.2 Multiple Radios per Platform

It might seem that the simplest solution to interoperability problems is to carry multiple radios. However, this is not a satisfactory solution to the interoperability problem because:

1. For data communications, complex interface equipment must be developed to provide connectivity between each of the radios and the onboard computers. The alternative, reconfigu-

---

[30]The SINCGARS radio is a specific example (one of many that could be cited) where interoperability with allied forces is possible only when the radio is used in its narrow-band mode: i.e., one cannot use the radio's ECCM/frequency hopping mode when interoperability is required (*Jane's Military Communications*, 1998, p. 115).

ration of the aircraft comm suite for each mission, is time consuming because of the extensive testing that is required.

2. For foot soldiers, the weight of a single radio is probably all that they can handle. For ground vehicles, helicopters, and other platforms, limitations on weight, space, and electrical power consumption are important. In most cases, the addition of new equipment requires the removal of something else.
3. Operators must be trained on every one of the equipments. Radio functions, setup procedures, and panel layouts are generally different from one family of radios to the next.
4. Equipments operating in the same or adjacent bands (or harmonically related bands) may interfere with one another unless carefully shielded and filtered.

Equipment power and antenna requirements can also be problematic. Radios operating in the same general band, but not at the same frequencies, may be able to share an antenna; signals can be routed to and from the antenna using a diplexer containing filters that separate the frequencies. If one must be able to support simultaneous reception on one radio and transmission from the other, then a duplexer is required to prevent transmitted signals from one radio leaking into the front-end of the other.

2.5.3 Translating Gateways

Another approach to radio interoperability, and a central element of the Army's Tactical Internet architecture, involves the use of translating gateways. The Tactical Internet consists of EPLRS and SINCGARS nets, as well as gateways that link ("internet") these nets together. Nongateway nodes (EPLRS or SINCGARS radios) can talk to other radios (of the same type) in their own network without going through a gateway, network topology permitting. A connection to the other type of radio, or to a radio of the same type within another net, must go through a gateway. Each SINCGARS/EPLRS gateway consists of a SINCGARS radio, an EPLRS radio, and an additional piece of interface hardware, unique to the Tactical Internet, called an InterNet Controller (INC). According to Keller (1996), one version of the INC, which is made by ITT, is a card that can be installed in a SINCGARS radio (the "EPLRS ties into the SINCGARS INC with a cable connector"); another version (in development at that time) is a standalone box. The INC "acts as both a bridge and a gateway to the U.S. Army's other legacy systems, with automated interfaces to EPLRS and MSE. . . . the INC converts and matches protocols across a wide variety of tactical command and control systems, including MSE" (*Jane's Military Communications,* 1998, p. 115). The INC also functions as an IP (Internet Protocol) router.

The Tactical Internet can be thought of as a collection of homogenous EPLRS and SINCGARS subnetworks that have been stitched together with translating gateways. This type of network architecture involves a number of drawbacks relative to a more homogenous network:

- the expense of additional hardware.
- more complex network protocol design.
- extra cabling, and a more complex network configuration.
- the potential for increased delays because of the constraint that traffic between incompatible nodes must pass through a translating gateway.

- constraints on platform movement that may be operationally unacceptable. If an EPLRS net needs to maintain connectivity to a SINCGARS net, then both must move so as to maintain line-of-sight connectivity to a common gateway.
- decreased network reliability/survivability because the gateway is a single point of failure. Failure or destruction of a gateway might partition the network into isolated subnetworks unless a backup gateway is available.

Translating gateways can provide interoperability between radios that use the same spectral bands under some conditions, e.g., one or both systems use suitable spread spectrum waveforms. However, noninteroperable radios that use the same spectrum may create unacceptable levels of mutual interference when operating in close proximity (e.g., within a few kilometers), requiring time-consuming coordination on frequency selection for combined operations.

2.5.4 Software Radios

In any radio, part of the processing must be performed using analog components (e.g., the low-noise amplifier, front-end filtering, and mixing). However, new technology permits more of the processing to be performed digitally, with the analog-to-digital conversion moving closer to the front-end (the limit of this is sampling directly at the output of the low-noise amplifier). In principle, any processing that is done digitally can be performed using software if the microprocessor is fast enough. Certain operations that cannot be performed rapidly enough in software can be handled via field programmable gate arrays (FPGAs). The term *software radio* refers to a new type of radio that is reprogrammable via a combination of software and FPGAs.[31] Software radios hold out the promise of several important benefits:[32]

1. Forward compatibility with new waveforms, including modulation and error control coding techniques that haven't even been invented yet, since the radio can be reprogrammed.
2. A reduction in the number of components.
3. The ability for one piece of hardware to have multiple "personalities." For mission 1, one might load personality X (this might be done at the depot) because low probability of detection is needed for that mission, while for mission 2, one might load personality Y in order to be able to communicate with units using a particular type of legacy hardware radio.
4. The ability to switch between different radio emulations "on the fly" in order to provide interoperability with any legacy equipments (hardware radios) that happen to be in the vicinity.

---

[31]This concept is an outgrowth of the Army Speakeasy program.

[32]In addition to the benefits listed below, software radios offer some benefits that are simply a consequence of doing more of the processing digitally. These same benefits would be realized using digital "hardware" radios. For example, digital filtering offers several advantages over analog filtering, including zero unit-to-unit variation, and zero degradation/change as the unit ages.

One of the key problems for software radios is the limited sampling rates of available A/D converters (about 40 million samples per second at 12-bit resolution). This sampling rate limits one to fairly narrow bandwidths except at VHF and UHF. Also, the power consumption of the A/D converter, which is the same whether one is transmitting, receiving, or waiting to receive, may be unacceptable for handheld equipments with limited battery weight.

5. Reduced platform radio footprint and power savings in situations where two (or more) conventional radios can be replaced by a single software radio that can emulate both.

The DoD plans to develop a software radio called the Joint Tactical Radio (JTR), which will offer much higher throughputs than existing radios and also be capable of interoperating with JTIDS and EPLRS (*C4I News,* April 10, 1997). Because the number of potential problem interfaces is roughly proportional to the square of the number of different radios fielded, phasing out of some legacy equipments would tend to reduce the number of interoperability problems and could be an important ingredient of an overall solution.

3. Alternative Architectures for Future Military Mobile Networks

In this chapter we explore a few of the more interesting architectural concepts for future military mobile networks and compare them qualitatively.

3.1 Rapidly-deployable Cellular Networks

There is considerable interest within the military in the tactical use of cellular telephones with modified base stations that are transportable (a system produced by Ericsson[33] was demonstrated at the JWID '97 exercise). Disaster relief operations and certain types of law-enforcement operations might also benefit from such systems. Tactical use of cellular telephone networks with transportable base stations may be practical for rear areas. However, there are severe drawbacks associated with the tactical use of unmodified cellular systems in forward areas:

- Base stations must be carefully sited, typically on high ground, and most are not transportable, i.e., once installed, they cannot be moved. In order to guarantee good coverage, commercial cellular service providers make extensive measurements as part of the base station siting process in order to minimize the incidence of coverage "holes."
- Commercial cell phones operate in fixed frequency bands, i.e., they do not have replaceable frequency elements (crystal oscillators and filters), and frequencies of operation cannot be changed through other means. Although GSM cell phones are produced in several versions, with the version for a given country designed to use the frequencies allocated for GSM in that country, even the GSM cell phones cannot be modified after production to adjust the operating frequencies. This is a major problem for the military, which needs to have common equipments that can be used wherever they may have to fight; such equipment must be capable of being adjusted to accommodate the frequency allocations of any host country.
- With existing rapidly deployable cellular systems such as the Ericsson system mentioned above, it must be possible to cover the area of operations with a single base station. (In a commercial system with multiple base stations, the base stations are interconnected via either land lines or microwave line-of-sight radio. Setting up these interbase station links is probably inconsistent with rapid deployability). Coverage radii of cellular base stations depend on terrain, antenna heights, power levels, and other factors. In general, however, radii larger than about 5 miles cannot be covered with a single base station. When multiple base stations are used, the stations must be interconnected to provide full connectivity among all users. A suitable solution to the problem of rapidly configurable links between transportable base stations has not yet been found, but it might involve either microwave or optical links.
- Users must maintain proximity to the base stations in order to maintain connectivity.
- Commercial cellular waveforms do not provide for jam resistance or LPD.
- Except for cellular systems that use CDMA, frequency planning is required to avoid unacceptable interference between adjacent cells. Regardless of what waveform is used (FDMA,

---

[33]This system was called the Mobile Expeditionary Cellular Communications Site (MECCS). For further information on MECCS, go to the URL http://www.jwid97.bmpcoe.org/demo.html.

FDMA/TDMA, or CDMA), suboptimal placement of cells (likely when base stations must be sited in a hurry) will tend to result in coverage holes and reduced system capacity.

- Although there are software tools to assist the network designer with base station siting and frequency planning, the process is still time consuming and requires highly trained personnel.

There is unavoidable "down time" for transportable communications equipment (unlike fully mobile equipment) that occurs when the equipment is torn down, packed, moved, unpacked, and set up; this dead time is probably unacceptable for many types of operations. (Moving and assembling the equipment may also be labor intensive and require specially trained personnel.)

The Army believes that cellular PCS technology may have a place in the command post or theater operations center, providing short-range wireless subscriber access to replace wired telephones.

Tactical use of rapidly deployable versions of commercial cellular telephone networks appears to be inappropriate for most types of military operations.

Possible exceptions include:

- use in the command post or theater operations center.
- peace keeping operations in and around urban areas.
- certain types of operations involving special operations forces, especially when the entire area of interest can be covered by a single base station, and where LPD is not a requirement.

3.2 Mobile Mesh Networks

Mobile mesh networks were an outgrowth of the earlier packet radio network concept.[34] The goal of packet radio network networks is to provide robust point-to-point communications for such connectionless, non-real-time traffic as text messages and noninteractive voice (voice messaging). Information is divided into packets which are then delivered in a store-and-forward fashion via one or multiple hops. Consecutive packets associated with a given message can take different routes through the network. For messages that must be delivered quickly and also with high reliability, associated packets might possibly be duplicated, with different copies following different routes through the network.

By the time packet radio network researchers had devised good protocols for supporting point-to-point non-real-time traffic, the needs and expectations of military users had changed. In particular, support for secure, efficient multicast (one-to-many and many-to-many) traffic and support for connection-oriented traffic with real-time requirements such as interactive voice and video had also become requirements. The term "mobile mesh network" has come to imply support for these additional services.

---

[34]See Bertsekas and Gallager (1992) for a general overview of the packet radio network concept. The survey issue of the *IEEE Proceedings* edited by Leiner et al. (1987) and the article by Kahn (1978), although somewhat dated, are also useful.

Corson and Macker (1996) present some strong arguments in favor of the mobile mesh network architecture for forward areas:

> In rear combat areas, communications may be facilitated through rapidly-deployable cellular infrastructures. However, in frontline areas, or during amphibious operations or rapid mechanized advances, battlefield conditions are often chaotic. Highly mobile units, such as artillery or tanks, quickly detach from combat groups to join and support others. Within these groups, the networking technology supporting their communications must be instantly reconfigurable, yet completely decentralized, redundant and survivable—in short, a mobile wireless mesh. A cluster of ships and airplanes in a naval task force also forms a mobile mesh. During amphibious operations, this cluster must communicate with mobile land-based units as well.

Although one can conceive of mobile mesh networks with several types of nodes differing in transmitter power levels, numbers and types of antennas, buffer capacity, processing power, and other characteristics, the design of the network protocols would be greatly complicated if one tried to take advantage of the special features of every type of node. If one did not take advantage of the special features, but relied instead on the minimum capabilities common to all nodes, then the special features of the more capable nodes would be wasted. Thus, homogeneous mobile mesh networks appear to have the potential for greater efficiency (at modest complexity) than heterogeneous ones. Because of the "curse" of legacy equipment, however, near-term Army mobile mesh networks will (like the Tactical Internet) be heterogeneous.

A closely related concept is that of the *self-organizing hierarchical network*. These networks are similar to the homogeneous mobile mesh networks, except that nodes organize themselves into clusters and by some means "elect" a cluster head (see, for example, Alwan et. al, 1996). The cluster head is responsible for keeping track of the membership of the cluster and the locations of nearby cluster heads, and for performing routing, switching, and trunking functions. However, since any node must be able to function as the cluster head, there is no savings of hardware relative to the mobile mesh network. The Near Term Digital Radio (NTDR), which is being built for Army experimentation, will function as a self-organizing hierarchical network (see Ruppe et al., 1997). NTDR is targeted to support TOC-to-TOC communications (TOC is Tactical Operational Center).

The obvious drawback of the mobile mesh architecture is the extra complexity that must be present at every node in order to keep track of network state information and route and buffer packets. Mobile mesh networks will have to be more costly than (non-self-organizing) hierarchical networks with mobile backbones because all nodes in a peer-to-peer network must be able to buffer multiple packet streams, route packets, and store information about the state of the network. For a Navy network in which every node is a destroyer-class or larger ship, the mobile mesh network concept seems appropriate, since the cost of the communications equipment is small compared to the value of the platform. However, it does not appear sensible to put a router on the back of every foot soldier (or on every ground vehicle) that needs communications capability. Because of constraints on power consumption and complexity, handheld radios with multiple receivers and the ability to route and buffer multiple packet streams will probably not be technologically feasible (except at low data rates) for more than a decade. The author believes that manpack-sized equipments with these capabilities could be developed in the near term with a significant R&D investment.

Providing connectivity into wired networks via a single wireless hop is a comparatively easy and cheap way to support mobile users. In the nonmilitary world, and for military bases and rear areas, most mobile users who need or want wireless communications operate in the vicinity of wired networks.[35] Thus, the only real need for mobile mesh networking is for the mobile warfighter in forward areas, and for certain types of disaster relief and police operations. The commercial world is unlikely to develop entirely new types of wireless networks, including hardware and protocols, for this limited market.

There is a further problem with the mobile mesh network concept: to achieve adequate connectivity it might be necessary to require equipments to be on continuously, which drains batteries quickly. One suspects that dismounted soldiers would surreptitiously turn off their radios when not using them in order to conserve battery power. ("Why run my batteries down so that my system can act as a repeater for someone else?") A more important concern, however, might be covertness. A user might not want his radio to be transmitting without his control because these transmissions might give away his position to the enemy.

For a discussion of unicast routing protocols for mobile mesh networks, we refer the reader to Johnson and Maltz (1996).

3.3 Fully Mobile Hierarchical Ground Networks

Fully mobile hierarchical ground networks are an interesting alternative to the mobile mesh network concept. Although less ambitious in terms of protocol and hardware complexity, some hierarchical network schemes might offer functionality comparable to that of the mobile mesh networks. In a fully mobile hierarchical network, the vast majority of the nodes would be relatively simple, e.g., with no special processing power and minimal buffer capacity. Only a small subset of the nodes (probably between 1 in 20 and 1 in 200) would have the necessary hardware to perform such functions as routing, switching, buffering of multiple packet streams, and trunking; we will refer to these nodes as *mobile base stations*.

The U.S. Army CECOM is developing a *radio access point* (RAP) that looks in many respects like a mobile base station. The RAP is "a HMMWV mounted assemblage of transmission and switching equipment that supports voice, data, and video users in the brigade and forward area. It provides both local switching and network access to the ATM backbone, as well as direct support for users on the move. The RAP will support all of the Army 21 requirements for both user and network mobility, network survivability, and security with the RAP itself on the move" (Bateman and Graff, 1996).[36] For operations in adverse terrain, or operations involving dismounted soldiers only, e.g., special operations such as sabotage, other solutions are needed.

A mobile base station, like a fixed base station in conventional cellular networks, provides connectivity to the users in its immediate vicinity. In addition, these base stations would have to be

---

[35]Airline travelers who use laptop computers are an important exception.

[36]An anonymous reviewer made the following comment: "While the RAP appears like a mobile base station its use on the battlefield is different. It is an entry point to higher echelon ATM networks where the mobile subscribers below the RAP are generally thought to consist of mobile mesh (or hierarchical) networks. The network is definitely multihop below the RAP. A base station generally implies cellular (i.e., direct) connectivity to the base station."

interconnected (crosslinked) in some fashion. The Space and Terrestrial Communications Directorate of CECOM initiated a program in spring of 1997 to develop a high-capacity trunk radio (HCTR). The objective capability includes trunking rates of 1.544 Mbps (T1) to 155.52 Mbps over distances of up to 40 kilometers (22 miles).

It is important to note that the traffic loading patterns are likely to be quite different from the patterns in commercial cellular networks. In commercial cellular networks, most of the traffic is mobile-to-fixed or vice versa, i.e., there is very little intracellular traffic. Thus, the ratio of required interbase station link capacity to user-base station capacity is fairly large. In the mobile tactical network, on the other hand, it seems likely that the vast majority of traffic would be intra-cellular, although this would depend on cell sizes, unit dispersion, and other factors. Certain kinds of network use could put a heavy load on the interbase station crosslinks. Examples include extensive use of the network by forward units calling for artillery fire from batteries in cells further back, or (better yet) forward observers sending real-time video back to the division command post.

In this section we discussed the concept of a hierarchical fully mobile network in which only a small fraction of the mobile nodes need to be capable of performing routing, switching, buffering of streams en route to other nodes, and longer-range communications. It is our opinion that this approach looks more practical and cost-effective than the hierarchical and mobile mesh network concepts that the Army is presently pursuing, in which all nodes are capable of performing these functions. Further analysis, simulation, and testing must be done to establish the best architecture for future Army mobile networks.

In hilly or mountainous terrain, connectivity will be a major problem for a system in which all of the mobile base station nodes are on the ground. Network connectivity based on a mesh of ground line-of-sight links requires that relays be located on high ground. "When these critical relay sites must be fortified and defended, support requirements can consume 7 to 8 percent of combat manpower…" (Rhea, 1997). This was the situation in Bosnia last year.

Future Army maneuver concepts call for significant blurring of the demarcation between friendly area and enemy area, so that the concept of a FLOT may no longer apply. This has important implications for the choice of network architecture, because enemy held or controlled areas between friendly controlled areas might preclude connectivity based on terrestrial relay paths alone. One possible solution to this problem is to use a combination of terrestrial relays and satellite relays. Another alternative is described in the next subsection.

3.4 Airborne Relay Concepts

One solution to the connectivity problem of the fully mobile hierarchical ground network is to put some or all of the base stations on airborne platforms. We begin by briefly discussing the full-mesh Joint Tactical Information Distribution System (JTIDS) architecture, since this existing system provides a useful reference point. In the remainder of Subsection 3.4, we describe various network architectures that use airborne relays (base stations) exclusively. Note that airborne relays will in general be more expensive to procure and to operate than ground base stations. Each airborne relay, however, can provide coverage to a larger area, so fewer will be needed. It is far from obvious what combination of ground based and airborne base stations represents the optimal choice, and we do not attempt to answer this question. Links to the airborne relay, involving much shorter ranges than a satellite link, require lower antenna gains.

3.4.1 Full-Mesh Networks vs. Relay Architectures

An Existing Full-Mesh Wireless Network: JTIDS

JTIDS is a communications system that was developed for air-to-air and air-to-ground communications. JTIDS multiaccess is based on radio nets; each of up to about 30 nets uses a distinct frequency hopping pattern. Each net is shared on a TDMA basis using 7.8125 ms slots. The key JTIDS feature of interest to us is that the network architecture is full mesh, i.e., all communications are via direct one-to-all broadcast (single-hop).[37] The main advantage of such an architecture is that there are no critical nodes, i.e., communications connectivity is not affected by the loss of any node.

Full-Mesh Networks Are Inefficient

Full-mesh wireless networks like JTIDS are inherently inefficient because one cannot make range (timing) and Doppler corrections at the transmitters, and because there is no frequency reuse. With a repeater-based architecture, all transmitters can adjust timing and frequency to correct for their range from the repeater and for relative velocity. In a full-mesh network, all of the other nodes are potential receivers, but one can make range and Doppler corrections for only one of them. With multiple repeaters (base stations), two repeaters that are not close to one another can use the same frequencies without interference; such frequency reuse enables large increases in system capacity over full-mesh and single-repeater architectures.

In a full-mesh network based on TDMA, required guard times depend on the maximum propagation delay from one end of the coverage area to the other. The guard time must be large enough that a transmission from a distant user in slot k does not overlap a transmission from a nearby user in slot k+1. JTIDS allows for two possible values of the guard time, depending on whether the maximum diameter of the network is chosen to be 300 nm or 500 nm. In the 300 nm mode, the guard time is

$$\tau_{300} = \frac{300 \text{ nm}}{3 \times 10^8 \text{m / sec}} \cdot 1.852 \text{ km / nm} = 1.852 \text{ ms},$$

which represents 23.7 percent of a 7.8125 ms slot. In the 500 nm mode, the guard time is τ_{500} = 3.087 m sec, which is 39.5 percent of a slot.

Efficiency Issues for Repeater-Based Networks

With a repeater architecture, guard times can be much smaller than in a full-mesh network. Transmitters can adjust their transmit times to correct for range differences, and guard times need

[37]JTIDS has a relay capability to support beyond-line-of-sight connections, but setting up a relay requires manual configuration by an operator, and the use of the relays also significantly degrades overall system performance.

be no longer than the uncertainty in the propagation time to the repeater plus timing errors. With accurate position and time information available at all nodes (from GPS, or from GPS in combination with an onboard clock), it should be possible to virtually eliminate guard-time overhead.

With a satellite or any other repeater, frequencies (or codes in a code-division multiple access system) must be divided between the uplink and downlink. In the typical satellite system, uplink and downlink bandwidths are equal, so that 50 percent of the bandwidth is "wasted" on the downlink. However, it is likely that the digital battlefield traffic mix will have a high percentage of short data transmissions, which are best handled via contention. With contention protocols like slotted Aloha, one needs only roughly a third as much capacity on the downlink as on the uplink, so that only 25 percent of the bandwidth is needed for the downlink.

3.4.2 Option 1: A Wideband Airborne Relay on an Existing Manned Platform

A repeater-based architecture could be implemented by placing a wideband airborne communications relay on an existing manned C^2 or SIGINT aircraft. The advantage of this approach is that no new platforms are required. There are, however, two major drawbacks:

- One may need to deploy this platform when its deployment might otherwise be unnecessary.
- For missions over hilly or mountainous terrain, it may be necessary to put the relay over the area of operations to provide line-of-sight connectivity to low-flying aircraft and ground forces. It is, however, undesirable to put a high-value manned platform in harm's way.

3.4.3 Option 2: A Wideband Airborne Relay on a High-endurance UAV

A wideband communications relay on a low-observable, high-endurance UAV (see Sass, 1997) addresses the two chief problems with Option 1. Multiple access protocols that have been developed to support a mix of traffic (short time-critical messages, interactive voice, file transfers, and other types) for satellite channels can be readily adapted to a UAV relay. If information about the UAV orbit is periodically transmitted to all participants via a secure channel, then timing and Doppler can be corrected at the transmitter in order to permit small guard times and close stacking of channels. By moving up from UHF to X-band or higher frequencies, one might be able to use as much as 500 MHz of spectrum for such a system. The net effect of all these factors might be a 30-fold increase in spectral efficiency relative to a system like JTIDS, and a 100-fold increase in system capacity. With multiple relays and frequency or code reuse, even larger capacities might be realized.

To minimize system vulnerabilities, it would be necessary either to have spare relays or to have a fallback mode that allows for direct communication without the relay. Although the communications hardware (in both the relay and user terminals), as well as the frequency assignments and protocols, are simplest for a single-repeater network in which all communications take place through the relay, this architecture has some significant limitations that lead one to consider enhancements of the basic concept. We briefly mention two potential enhancements.

3.4.4 Option 3: A Dual-band System with both Direct and Relayed Connections

A dual-band system with both single-hop (point-to-point) and double-hop (through-the-relay) connections allows for direct point-to-point links that bypass the UAV relay for short-range connections. These short-range links would use frequencies on the slope of the oxygen absorption band at 60 GHz; the strong attenuation limits these transmissions from interfering with other simultaneous transmissions occurring elsewhere at the same frequency. One can simultaneously minimize the risks of jamming and detection, while achieving the desired bit error rate at the intended receiver, by sliding away from the peak absorption frequency until the desired link quality is achieved (see Section 6.3.2). Longer-range communications would take place through the UAV relay. This approach offloads a significant part of the communications traffic from the relay, increasing the total amount of traffic that can be handled. Other benefits include reduced power requirements for short-range connections, improved survivability against ECM (it is more difficult for the enemy to jam or detect the 60-GHz band communications), and better connectivity between nonrelay aircraft and participants on the ground (important for such missions as close air support). A variation on this concept involves putting dual-band receivers on the UAV so that it can receive transmissions in the 60-GHz band from nearby sources; these transmissions would not be detectable to enemy ESM equipment except at very close range.

3.4.5 Option 4: An Airborne Backbone

A ring (or other biconnected network) of UAVs forming a "backbone in the sky," with UAVs netted via high-rate crosslinks, could be used to cover larger areas than could be covered with a single UAV. Implementing the high-rate crosslinks between UAVs requires an additional two antennas (possibly three for some non-ring topologies) on each UAV.[38] This concept is attractive because at least two UAV relays must fail before connectivity is degraded. Also, with sectorized antennas and frequency or code reuse, it should be possible to provide sufficient capacity to support real-time imagery and video. Note that option 4 is significantly more complex and expensive than option 2 (the single-UAV architecture), primarily because of the crosslinks, but also because the protocols must support handoff, i.e., as mobile users (ground vehicles or aircraft) and the UAV move, it will sometimes be necessary to switch a user's connection from one relay to another; this is only a problem for connection-oriented traffic such as two-way voice and real-time video. Requirements that would tend to drive the design in this direction include: (1) the ability to cover a larger area than can be covered by a single UAV, (2) the need for high capacity, (3) reduced EIRP requirements for the ground terminals, (4) reduced vulnerability to jamming, and (5) increased network survivability.

Options 3 and 4 could be combined. Further research is needed in order to explore these and other alternative concepts in greater detail, including both cost comparisons and performance in realistic scenarios.

---

[38]In order to achieve high crosslink data rates, e.g., on the order of 40 Mbps, the crosslink antennas must have fairly high gains and be accurately pointed. It is unclear whether a conventional antenna or phased array antenna should be used for the crosslinks; in either case, a variety of technical challenges exist.

4. Commercial Wireless Technology: Products, Standards, and Services

4.1 Introduction

Recall that the first goal of this study was to assess current and near-term commercial wireless technology, including both standards and products, and to identify the extent to which Army tactical communications needs are likely to be satisfied by them. We use the term "needs" rather than "requirements," since no formal requirements documents were consulted, and because our assessment is largely qualitative.

We note at the outset that conclusions that might apply to the problem of military use of commercial computer hardware do not apply to commercial wireless communications hardware. Aside from the common requirement for ruggedization, the two situations are very different; military and commercial users of computer systems have very similar requirements, while military and commercial users of wireless communications systems have very different requirements.[39]

For purposes of this study, it is useful to divide commercial wireless technology into four general categories:

- Components and subsystems.
- Waveforms and signal processing.
- Middle-layer protocols.
- Products and services.

In the subsections that follow, we examine each of the four areas and draw some general conclusions about the extent to which Army wireless communications needs can be satisfied.

4.2 Components and Subsystems

Components and subsystems are hardware items that are not usable except when integrated with other components and subsystems to make a complete system. Relevant component types include:

- application specific integrated circuits (ASICs) such as error control encoders, error control decoders, phase locked loops (PLLs), digital filters, and modem chips.
- general-purpose integrated circuits such as memory chips, microprocessors, and field programmable gate arrays (FPGAs).

---

[39]Real-time computing requirements tend to be more important for military applications than for commercial applications, but the impact of this requirement is primarily on the operating system and other software, rather than on the hardware itself.

- RF components such as waveguides, cables, high-power amplifier tubes (traveling wave tubes and klystrons), and ferrite devices.
- batteries.
- antennas.

Subsystems include such things as low-noise (low-power) amplifiers, high-power amplifiers, and upconverters and downconverters. Antennas are also sometimes classified as subsystems.

Motivated by the need to reduce acquisition and O&M costs, the Federal Acquisition Streamlining Act of 1994 mandated the elimination of most military specifications and standards, and increased use of commercial off-the-shelf (COTS) equipment and parts by military contractors.[40] There is considerable disagreement as to which components and systems qualify as "commercial." Rhea (1996) cites four criteria that must all be satisfied for an integrated circuit to be termed commercial:

(1) it must be included in a manufacturer's catalog and offered for sale to any potential customer;

(2) it must have a unique identifying part number that all users recognize;

(3) it must be fabricated using standard processes and assembled and tested using standard methods and equipment; and

(4) it must be interchangeable with all other devices bearing the same part number and, where several grades are offered, be downward compatible with all lower grades of the same product.

In the present environment, military contractors have a strong incentive to call parts "commercial," regardless of whether this is justified. Note that criteria (1) and (2) are trivially satisfied, and (4) may also be trivially satisfied unless there are multiple grades of the product. Thus, the third criterion is the real touchstone of commercial status.

Because of the much larger production volumes for commercial components, as well as competition among multiple manufacturers and suppliers, costs tend to be substantially lower for commercial components. Concerns in some quarters that use of commercial-grade components would lead to reduced system reliability have tended, for the most part, to be unfounded. The notion of a two-way commercial/military divide is overly simplistic; commercial components are often available in a range of different grades; it often happens that some of these grades are better than MILSPEC grades. In fact, because of the larger-scale production, commercial components are often more reliable and exhibit less unit-to-unit variation than comparable MILSPEC components.

The decreased use of MILSPEC components has resulted in an exodus of suppliers from the military market, particularly among manufacturers of military chips.[41] For many types of components, there is no clear need for military-specific components. Consider the example of

---

[40]A more detailed discussion of this law can be found in a document at the following URL: http://www.pica.army.mil/orgs/fsac/aif_mo/acquisition.html.

[41]In 1994, there was speculation that the few gallium-arsenide chip foundries would go out of business because of their heavy dependence on military business; but because of a sharp upturn in the commercial market for high-speed VLSI chips, this did not happen.

batteries. Except possibly for the need to operate over a greater temperature range, the needs of mobile business people are very similar to those of the mobile warfighter—both need lighter-weight, more compact batteries that can withstand larger numbers of charge-discharge cycles. Thus, the demand of the commercial marketplace is sufficient in this case to promote R&D that, if successful, will satisfy both user communities. Thus, it would appear that there is no need for military-funded R&D in this area. A possible exception is "one-shot" (single-use, nonrechargeable) batteries offering very high energy densities; these are likely to remain of interest primarily for military applications because of their high cost.

In some technology areas, however, military-specific components will continue to be necessary, and the diminishing pool of suppliers in these areas gives cause for alarm.

Where there is insufficient commercial demand for a class of component or subsystem that the military needs, it may be worth taking aggressive measures to ensure the continued existence of a reasonable pool of both suppliers and R&D technical expertise.

Two component technology areas where some form of subsidy or other incentive for R&D may be advisable are (1) broadband high-power amplifiers (in particular, for frequencies at X-band and above) and (2) high-gain low-sidelobe antennas and other antennas with unusual characteristics.

Plastic encapsulated microcircuits (PEMs) are an interesting component technology area, because the picture here is clouded in controversy.[42] PEMs, which are the mass-produced microcircuits used in virtually all commercial electronics, tend to be much cheaper than MILSPEC parts, which are either encapsulated in ceramic or enclosed in sealed cans filled with an inert gas. The type of plastic used in older PEMs was slightly permeable to moisture, so that exposure to high-humidity environments could damage the components over a long period of time. However, it would appear that with newer plastics this is no longer a problem. Although much of the opposition to military use of PEMs is based on the supposed susceptibility to moisture, other deficiencies might be more significant. Some military wireless equipments, because of the need to operate in high ambient temperatures, require better heat dissipation than comparable commercial equipments; it is harder to provide adequate heat dissipation with plastic enclosures. Also, thermal shock can cause plastic-metal seals to fail, and repeated temperature cycling can eventually lift wire bonds. Lastly, reliability testing for PEMs tends to be expensive, except when amortized over a very large number of units.

Arguments over reliability involve, in part, a difference between military and commercial business practices. When a piece of consumer electronics fails, the most cost-effective course of action is often to simply throw it away and buy a replacement.

---

[42]Much of this discussion of PEMs is based on an August 1, 1997, discussion with William Coomler of the Aerospace Corporation.

The military needs to weigh the advantages of relatively inexpensive "disposable" equipment that is designed for shorter lifetimes. These advantages include reduced load on maintenance depots, and a reduced time lag for the introduction of new technology. For those operations where high availability is essential, the most practical way to ensure availability might be for each company to carry an extra radio or two. Designing for long mean times between failure (MTBFs) under extreme conditions results in high costs that are hard to justify.[43]

Widespread use of short-lifetime equipment might require a more timely resupply capability than currently exists.[44]

4.3 Waveforms and Signal Processing

Waveforms and signal processing correspond to the physical layer (layer 1) in the ISO Open Systems Interconnection (OSI) model. Elements of a waveform include:

- the basic waveform type: TDMA, FDMA, direct-sequence CDMA, frequency hopping, or other
- the baseband modulation: phase-shift keying, frequency-shift keying, etc.
- error control coding and interleaving (if any): convolutional coding, block coding, or a combination of the two
- synchronization preamble (if any)

In signal environments that are hostile, or simply crowded with competing users and interference from other systems, wireless link performance may degrade unacceptably. A variety of advanced signal processing techniques will be increasingly important for military wireless networks; these include digital demodulation, jammer side information (JSI) processing, adaptive filtering for interference rejection, adaptive equalization, array signal processing, and multiuser detection. Some of these techniques require processing power that is not yet practical for handheld or other small terminals; performance of the hardware and algorithms may severely limit the rates at which information can be transmitted, requiring a tradeoff between link quality/robustness and user data rate. With improvements in chip densities, power requirements, and algorithms, the use of many of these techniques is already practical for small terminals, or will become practical in the near future.

There is a strong divergence between military needs and what the commercial world is providing and likely to provide in the near future. Primary reasons for this divergence include:

1. Different priorities for commercial and military wireless system users. Advanced signal processing generally requires more complexity, and this in turn increases equipment costs and power requirements. Increased power requirements translate into larger and more

---

[43]These arguments apply primarily to ground systems; the picture of avionics and satellite systems is different. High integration costs, long replacement lead times, and for satellites, high bus and launch costs all drive the design toward highly reliable communications payloads.

[44]Several recent RAND reports on military logistics address this and other related problems.

expensive transmitters and increased battery weights. Cost, power requirements, and weight are important for military equipment, especially when one deals with manpack equipments and equipments for special forces. However, military users are in general willing to accept somewhat greater weight in return for critical functionality such as increased A/J and LPD.

2. In the commercial world, relatively small differences in user equipment costs or service costs can determine which of two competing systems is ultimately successful in the marketplace. In the acquisition of military wireless systems, performance, as reflected in system requirements, plays a more important role. If performance requirements cannot be satisfied at reasonable cost, then it may be necessary to re-examine the requirements, but cost is not (or should not be) the primary driver in the design of military systems.

4.4 Middle-layer Protocols: Next Generation IP Protocols

4.4.1 Introduction

The middle-layer protocols are the OSI link, network, and transport layers (layers 2–4). These protocols are typically implemented using a combination of software and firmware, although some link layer functions are implemented directly in hardware. Layers 3 and 4 include the TCP/IP Internet protocol suite used in the Internet and in military IP-routed networks such as the SIPRNET (Secret-level IP-Routed NETwork). Compatibility of mobile military networks with these IP-routed networks would be very useful. The easiest way to achieve such compatibility is by using the same layer 3 and 4 protocols in the mobile wireless networks. However, requirements for mobility and mobile multicast may force the military to use middle-layer protocols that are incompatible with the TCP/IP protocol suite. The remainder of Subsection 4.4 is devoted to these issues.

4.4.2 IP Mobility: Mobile IP and IPv6 Mobility

The current version of IP is version 4; the next will be version 6 (version 5 was skipped). Mobile IP is a proposed addition to IPv4 that would address some mobility issues during the transitional period until the IPv6 standard is ratified and the protocol becomes widely available on routers. Because mobile IP and IPv6 mobility are similar, although not identical, we shall use the term *IP mobility* to refer to both.

> IP mobility enables a very limited type of mobility in which mobile hosts are permitted, but not mobile routers; two mobile hosts can thus communicate only if each is no more than one wireless hop away from the same fixed network. Core functions such as routing are performed within the fixed, wired part of the network (Corson and Macker, 1996).

IP mobility was designed so as to avoid the need for global changes to existing Internet routing; IETF working group members considered this to be an absolute constraint on the solution space. Mobility "agents" will be added to selected routers that need to support mobility; other routers do not need to be "mobile-aware."[45] We refer the reader to Johnson and Maltz (1996) for a more detailed discussion.

---

[45]The Internet Engineering Task Force (IETF) has, for obvious reasons, required that any changes to Inter-

In order to minimize the impact on applications, IP mobility is being implemented in a fashion that is transparent to applications. In particular, mobile IP hides information about changes in the network environment from higher protocol layers. This approach offers the benefit that existing network applications will not require changes, but it also has at least two major drawbacks:

1. Data rates and error rates of mobile wireless connections vary over time; this variability will be more severe in military networks with mobile backbones than in commercial networks with fixed, carefully sited base stations. For wireless connections with real-time requirements, the mobile military user needs applications that can respond to changes of bandwidth and error rate. For example, for interactive speech, one might want to use available data rate information to control vocoder output rates for digital voice to match the data rate of the channel.

2. The military is providing intelligence and other information over the IP-routed networks such as SIPRNET; this information will increasingly be accessed via standard Web browsers such as Netscape. Proxy Web servers need to know what type of environment a mobile node is in, so that when a user moves from a high-speed network to a low-speed network, multimedia information can be automatically sized to the link, e.g., by transmitting smaller versions of images, or if necessary, text-only versions of Web pages).

Protocol features that allow applications and users to make the most of whatever network resources are available at any time is an urgent requirement for the military. Commercial protocol standards are unlikely to incorporate such features because they are not a high priority for the commercial world, in part because connection characteristics are more predictable in networks with fixed base stations, and in part because lapses of connectivity are more acceptable for the user community that these standards are being designed to serve.

4.4.3 Multicast Routing

The military, and the Army in particular, need not only one-to-one communications, but also one-to-many and many-to-many ("netted") communications. With a combat net radio such as SINCGARS, users are assigned to nets, and all users on a given net hear each other's transmissions. In general, unless a user has multiple receivers, he cannot receive or transmit on more than one net at a time. (Netted communications require some mechanism to prevent multiple simultaneous communications on a given net.) In packet-switched networks, the analog of netted communications is called *multicasting*, or *multicast routing*. Multicasting is actually more versatile than netted comms because multicast groups may have overlapping membership, i.e., a given user may be a member of more than one multicast group.

Benefits of Multicasting

Although it might seem that multicasting is equivalent to a series of one-to-one communications, multicasting confers three important benefits:

---

net protocols be implemented in such a fashion that the new protocol software can be installed gradually on individual hosts and routers. Thus, problems associated with backward compatibility and legacy systems are not unique to the military.

1. More economical use of resources, including both link capacity and buffer capacity. This is particularly important for wireless networks with low data rate links.
2. Nearly simultaneous receipt by all receivers. This is particularly important for dissemination of command and control information in military networks.
3. The ability to support interactive group communications.

To see why multicast routing is more efficient than sequential one-to-one routing, consider the mobile hierarchical network shown in Figure 2. Circles and squares indicate mobile base stations and mobile terminals, respectively. Lines show which pairs of nodes can currently communicate. Suppose that node 1 needs to send a message to nodes 2-7. With sequential one-to-one routing, the same message is transmitted 6 times by node 1, and a total of 16 times counting all hops (link traversals). With multicast routing, node 1 transmits the message only once. The message is received by base station A, which then broadcasts to nodes 2 and 3 and transmits to base station B (two transmissions). Base station B then broadcasts to nodes 4-7 (an additional transmission). Thus, the total number of transmissions is reduced from 16 to 3. Note that because the interbase station links have comparatively high rates, all 6 recipients receive the message at nearly the same time when multicast routing is used.

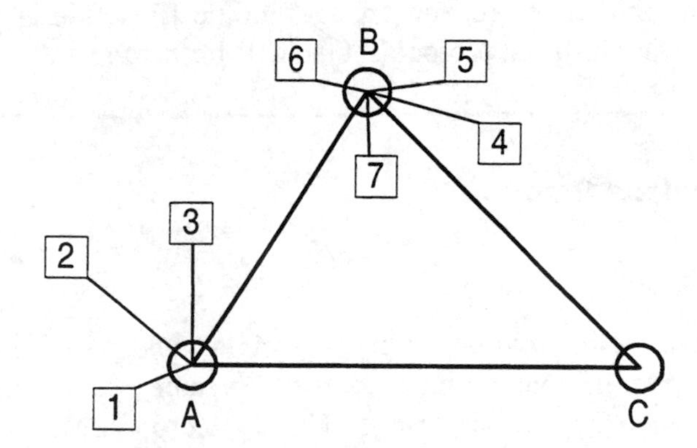

Figure 2—A hypothetical mobile hierarchical network

Shortcomings of Emerging Standards for Multicast

Multicast extensions of the TCP/IP protocol suite are being developed to support live multimedia distribution (e.g., for pay-per-view events) and also bulk file distribution (e.g., for software updates) to large numbers of users over the Internet and corporate intranets. However, the underlying algorithms depend on both the relatively static topologies and the reliable links in these wired networks.

Multicast routing will be essential for future military wireless networks. However, because commercial multicast routing protocols are being designed for wired networks with static topologies and reliable links, they will not be useable in fully mobile wireless networks.

The military also has special requirements for secure reliable multicast for multicast groups with dynamic membership. These requirements will also not be solved by commercial multicast standards.

4.4.4 Conclusions

IPv6 has numerous useful enhancements, e.g., a larger address space and support for encryption. However:

> The type of mobility support provided by mobile IP and IPv6 will not serve the Army's needs for mobile wireless networking. On the other hand, it will be difficult for the military to develop competing products because of the complexity of the protocols and the difficulties associated with testing.

Testing is particularly problematic, because one does not learn enough from small-scale tests and demonstrations, while large-scale testing is very expensive. (The testbed for IPv6, known as the 6Bone, involves more than 20,000 routers around the world.)

> It is important for the military to remain engaged in the IP standards process and to fund research to develop middle-layer protocols that will be better suited for military mobile wireless networks.

4.5 Commercial Wireless Products

4.5.1 Introduction

Because of the large cost differential between commercial- and military-developed equipments, there is a strong incentive for the military to make greater use of commercial off-the-shelf (COTS) equipment. Increased military use of COTS microprocessors for embedded control applications is an example of a COTS application with no apparent "downside." There are, however, several major drawbacks associated with the tactical use of unmodified commercial wireless equipment:

- As discussed before, commercial wireless systems are dependent on fixed (or in some special cases transportable) infrastructure, i.e., they are not fully mobile.
- Most of these equipments provide inadequate security.
- With the exception of some high-end equipments, frequency elements (crystal oscillators and filters) in most commercial wireless transmitters and receivers are not replaceable.[46] Without replaceable frequency elements, these equipments cannot operate on U.S. military frequency bands and on bands that are likely to be available in host nations overseas.
- Commercial bands vary from country to country, so that equipments that can be used in North America may not be usable in other countries because the appropriate supporting

[46]Some high-capacity microwave radio relays for telephone trunking applications are designed with replaceable frequency elements.

infrastructure is lacking. In most cases, one cannot simply fly in base stations and set them up because the requisite frequencies may be assigned to other systems in that country, or reserved for future use. Negotiation with host countries is often required.

- Most commercial equipments are insufficiently rugged, although after-production ruggedization is often possible.
- Commercial systems lack jam resistance.
- Commercial waveforms don't provide low probability of detection (LPD).

Commercial wireless equipments are generally unsuitable for most military operations. Use of wireless LANs, cordless phones, and other short-range systems at base and in rear areas are possible exceptions.

Although military variants of commercial systems would appear to offer a middle course of action between military-unique and commercial off-the-shelf systems, military variants tend to represent a poor compromise between these two extremes. Except for ruggedization, there is little that can be done on an after-production basis. Other modifications require the cooperation of the original equipment manufacturer. Several factors tend to drive up the costs of military variants of commercial equipments:

- Military variants require design elements and design practices that are unfamiliar to most commercial design engineers. This tends to increase the design costs.
- Chip designers who need to implement military-specific waveforms and protocols, or who must design for wide ranges of operating temperature, tend to be able to make less use of "IP" (intellectual property).[47]
- Because production volumes of military systems are relatively small, the impact of these higher design costs on per-unit costs tends to be large.
- Testing requirements are much more stringent for military equipment.
- Since relatively few companies are willing to produce military variants, there may not be much competition.

The result of these factors is that prices tend to be much higher than for unmodified commercial equipments, and may approach those of military-unique systems.

[47]In VLSI chip design, IP refers to prefabricated blocks of gates that perform commonly needed functions and are available for use by any designer subject to payment of royalties to the copyright holder.

4.5.2 Wireless LANs (WLANs) and IEEE 802.11

The IEEE has recently finalized a standard, designated 802.11, for wireless local area networks (WLANs), and the market for WLANs is expected to grow rapidly. The primary tactical application of commercial WLANs would be for computer-to-computer communications within command posts; set-up and tear-down times for command posts would be greatly reduced, since all computer LAN cabling is eliminated (power cables are still necessary).

The standard includes three alternative, incompatible physical layer interfaces. There is no provision for internetworking between different types of wireless LANs; a given hub ("access point") will work with only one type of wireless LAN.

One of the three approved physical interfaces uses a form of frequency hopping; prior to 802.11, frequency hopping was an exclusively military technology. Even more interesting, the network can operate either with or without a hub (without a hub, the network operation is peer-to-peer, but range is very limited). In order to simplify the process by which nodes join the network, the standard requires that network nodes periodically broadcast a beacon that contains the key for generating the hopping pattern. Availability of the hopping pattern eliminates any anti-jam benefits of frequency hopping.

> For many tactical applications, use of wireless LANs that conform to the IEEE 802.11 standard is unacceptable because of the potential for eavesdropping on unencrypted communications and for jamming.

4.6 Commercial Wireless Services

4.6.1 Introduction

Commercial mobile wireless services include:

- analog and digital cellular telephony.
- personal communications services (PCS).
- paging.
- wireless data services.
- interactive voice and low-rate data services over LEO satellites.
- messaging over LEO satellites (ORBCOMM offers a service of this type).
- high-rate data over LEO satellites (Teledesic plans to offer a service of this type).

A major concern about the military use of any of these systems is that the waveforms are not designed against jamming or to provide LPD. Furthermore, unlike military communications systems, detailed waveform characteristics are generally available. With the military systems, potential adversaries cannot begin to collect information about the system and develop countermeasures until the system is fielded and used. Developing jammers or direction-finding equipment that is optimized for use against the system can thus be delayed by several years. With standards-based systems, developers of the systems and developers of countermeasures both begin work at the same time.

4.6.2 Cellular Telephony and Personal Communications Services (PCS)

Cellular telephone systems now cover metropolitan and suburban areas in North America and Europe, as well as much of Asia and the Middle East. The Advanced Mobile Phone System (AMPS), an older, analog system deployed throughout much of the United States, is gradually being supplanted by three newer digital cellular standards. There is some confusion concerning the difference between cellular telephony and PCS. The term PCS seems most useful in reference to a class of digital wireless systems that provide two-way voice in combination with at least one other nonvoice service such as text messaging. Whereas cellular systems are designed to support both users in vehicles and pedestrians, and to offer nearly complete coverage of metropolitan areas and major highways, some PCS systems support only pedestrians within relatively limited areas such as train stations and shopping malls. Because of the similarities between these two types of systems, we treat them together.

At $300 or less, the cost of a cellular telephone handset is far lower than that of any military radio; the services are also typically modestly priced. On the other hand, tactical use of existing cellular telephone networks entails several potential problems:

- The military needs systems and services that can be used anywhere they may need to operate. For economic reasons, commercial cellular services, although proliferating, will only be available in populated areas and along major highways.
- These systems depend on wired infrastructure that might be destroyed by fighting or by a disaster, and that could be easily sabotaged.
- Because these systems do not recognize different classes of users, military users would compete with the public at large for access.
- Analog systems and most digital systems provide no mechanism for secure voice communications. Those digital cellular schemes that do provide for encryption use fairly short keys that could potentially be broken. Even more worrisome, there is no end-to-end encryption, i.e., only the over-the-air part of the connection is encrypted.
- Because different standards have been adopted in different areas (four types of cellular telephone networks exist in North America alone), one must have a substantial inventory of cellular phones in order to be able to use any cellular system.
- Although GSM cellular telephony networks have been deployed in some 60 countries, differing frequency allocations in different regions of the world prevent one from using a single phone in all of these different locations.

Tactical use of existing cellular telephone networks makes sense only for peace keeping and disaster relief operations in urban and suburban areas (assuming that the disaster has not rendered the system unusable).

4.6.3 Wireless Data Services

The main wireless data services currently in use in North America are RAM, ARDIS, and CDPD. Of these, CDPD appears to be the most successful. CDPD, a digital overlay on the analog AMPS cellular system, uses idle 30 kHz voice channels to send packetized data at a rate of 16 Kbps. Although one can send data over digital cellular systems (and even over AMPS, if one uses a suitable modem), digital wireless services are far more efficient for short messages (say, less than 2000 bytes) because they avoid the overhead associated with call setup and termination.

Wireless data systems are being used by package delivery companies, and CDPD is being used successfully by law enforcement agencies in some areas of the United States. Note that any portable computer equipped with a suitable wireless data modem and software drivers can have wireless access. Encryption can be easily implemented, since data can be encrypted via software before being transmitted. Unfortunately, almost all of the objections that we raised in connection with the military use of cellular services apply to the wireless data services.

4.6.4 LEO and MEO Satellite Services

There is considerable interest within the military in the use of commercial low Earth orbit (LEO) and medium Earth orbit (MEO) satellite services for interactive voice, low-rate data, and high-rate data. Market predictions vary wildly, but even the most optimistic predictions indicate that there will not be enough users to make all of the proposed ventures profitable; several must ultimately fail. As competing companies rush to launch their systems, it is important to bear in mind that, for each type of service, cost of the service is likely to be a more important indicator of long-term success than earliest availability.

Because all of these systems are incompatible, potential users are faced with the possibility of losing their investment in the terminal (handset) if the service provider goes bankrupt. DISA has procured (or is procuring) a gateway for use with Motorola's Iridium system; the government thus finds itself in the position of betting on the success of a system whose projected usage fees are 3-4 times higher than those of competing systems such as Globalstar.

For military users, there are some specific drawbacks of these systems that merit attention:

- Total capacities (measured in units of circuits for voice or bits per second for data) may look impressive but are fixed and relatively small for any given area of the world. Much of this capacity may eventually be taken up by other users who cannot be denied service because of a surge in military demand. Unlike geostationary satellites, which are at much higher altitudes, with LEO satellite constellations it is impossible to shift excess capacity from one region of the world to an adjacent region.
- Except for Iridium, all of these systems depend on regional gateways that might be attacked or destroyed by saboteurs in time of war.
- Because systems like Iridium and Globalstar have to negotiate "landing rights" with every sovereign country on an individual basis, it is possible that a host country where the U.S. military might be based would not have an agreement, in which case use of the handsets within that country would be illegal.

- Apparently, each satellite in these new systems will have a mechanism for geolocating a user handset in order to determine whether that user is transmitting from a country that has signed an agreement, so that landing fees (royalties) can be paid to the appropriate nation. A handset identification number and location information may be downlinked in the clear (i.e., without encryption) and thus be subject to interception; this is probably unacceptable from the standpoint of security.
- In order to minimize cost, high data rate systems like Teledesic will use ground terminals with gimbaled parabolic dish antennas rather than phased array antennas. The antenna drive mechanism is capable of tracking the satellite motion but cannot compensate for vehicle motion. Thus, in order to communicate, the user must stop moving.
- All of these systems use waveforms that are highly susceptible to jamming. For any of these systems, a fairly simple jammer could be constructed that could jam either selected channels or entire transponders.[48]

Conclusion: Most commercial LEO and MEO satellite systems provide only low rate data services; the military needs higher data rates. For high data rate systems such as Teledesic, the terminals will not be fully mobile. High-data-rate fully mobile communications could be achieved by building military-unique terminals with phased array antennas for use with Teledesic; however, the costs of such terminals would be high. For all of these LEO satellite systems, the financial viability of the service providers, as well as security and robustness against jamming, continue to be major concerns.

4.7 Government and DoD: To What Extent Can They Influence the Commercial World?

Increased military presence in the various standards working groups might help to influence the standards process. However, the needs of the military are in many cases radically different from those of the commercial world. Because the military market is comparatively small, the military can only hope to impact the standards process in areas where there are also nonmilitary applications for a given solution direction.

There is a general recognition that in most areas, the government cannot mandate standards, and that any attempt to do so is likely to be counterproductive. (Recall the failures of the GOSIP protocol suite and Ada.) Furthermore, the military is no longer a large enough customer to be able to determine what products win or lose in the marketplace. However, there are ways in which the military might be able to constructively influence the direction of commercial wireless. With regard to the standards development process, we suggest the following options:

- Participation of military user representatives in standards working groups and other forums. The primary value of such participation is to maintain a dialog and active exchange of information between the military and industry representatives in the working groups.

---

[48]Because the satellites are not geostationary, the jammer would have to use an antenna that tracks the satellite, with computer control of the drive system. While not a $400 "junk" jammer, such a jammer is also not very complex, and one could be readily assembled from commercially available components.

- Helping to promote research in areas that are of particular importance to the military, and where the commercial world seems unlikely to take the initiative on its own, e.g., because the market is primarily military, or because the risks are too high. Military research funding is not the only approach. Other techniques for incentivizing industry include joint ventures involving government subsidies for research carried out by industry, and guaranteed minimum buys of products meeting specified performance requirements.

It is often stated that the government cannot have much impact on the development of electronic products in the marketplace; however, there are some possible exceptions. For some specialty products, the commercial market is fairly small, and the military market may be significant compared to the commercial market in terms of numbers of units to be sold or net profit to be realized. In some cases, there is significant uncertainty among market analysts regarding the size of the market for certain products or services. This is currently true for some wireless data products and services. For some of these products and services, there is general agreement that a sufficiently large market *will* develop within the next decade or so; in the meantime, however, few companies are anxious to accept the risks associated with the development and introduction of new products, and some that have are losing money. In such cases, a small "push" from the government, e.g., in the form of a guaranteed minimum buy, or a promise to defray some part of the development costs (subject to performance requirements) might reduce the risk to the point where such ventures would be more acceptable. In such cases, the military might also have considerably more influence on the characteristics of the system.

5. The Channel Modeling Problem

The Army wants to transition to communications networks that are more mobile, capable of higher data rates, better integrated (smaller numbers of boxes with increased functionality), interoperable with a wide range of equipments (including legacy systems), easier to configure and use (including setup time), more survivable, and more affordable. Clearly, these goals are in conflict. Achieving a near-optimal set of compromises requires careful balancing of trades. Some aspects of performance can be assessed through simulation, while others (e.g., ease of use and reliability) require laboratory testing and user field trials.

Although simulation is an essential part of the process by which competing design alternatives are compared and inferior ones winnowed out, simulations of wireless communications links and networks are notoriously unreliable. For any type of stochastic system, there are at least four general ways in which simulation studies can lead to incorrect design choices:

A. errors in the underlying mathematical models of the system and its environment, in associated data, or in the process by which models are fitted to the available data. Errors in modeling include unwarranted approximations and simplifications.

B. faulty implementation of the models in code (e.g., programming errors or use of faulty random number generators).

C. improper inputs, e.g., inputs that violate the range of validity of the underlying models, or insufficient exploration of the parameter space.

D. errors in the statistical processing and interpretation of the simulation outputs. Examples include: (a) insufficient sample sizes (numbers of runs and/or run lengths), (b) failure to account for time dependence in simulation outputs, and (c) untested assumptions about the behavior of the model, e.g., that interactions between design parameters can be ignored.

Our focus here is on a specific aspect of A (errors in the underlying models)—the problem of characterizing the external environment in which a wireless system/network must operate. The external environment is the communications *channel* between a pair of antennas. The channel accounts for propagation effects such as ordinary $1/R^2$ free space loss, rain absorption, multipath, diffraction, refraction, and scattering, as well as general background noise. In the widest sense, the channel may also account for sources of interference when these are treated in aggregate. In any case, sources of interference, whether friendly (unintentional) or jammers, are part of the external environment.

The specific propagation effects that must be accounted for in a channel model depend on the type of system (including the frequency of operation; symbol rate, modulation, coding, and other waveform characteristics; antenna types; and antenna heights), the terrain, rates of movement, and other geometrical factors (e.g., distances between antennas and distances to reflective surfaces). Diffraction, refraction, and scattering are often of secondary importance and can be ignored for many applications.

Multipath, however, is an important effect that is often disregarded or improperly modeled. For low-data-rate systems, multipath can often be represented as flat fading, i.e., as a time-varying attenuation that affects the amplitudes but not the shapes of received signal pulses. For higher-data-rate systems, however, multipath causes distortion of pulses and intersymbol interference; these effects can be crucial in determining waveform parameters and receiver characteristics, including equalization, rake reception, error control coding and interleaving, and the use of spread spectrum techniques.

There are at least four major problems associated with the channel models used in wireless communications system simulations:

1. There is a lack of standard reference channel models that can be used for making fair comparisons between competing system concepts. Contractors and other proponents of systems are essentially free to choose the external environment models against which their systems will be evaluated. There is a strong disincentive to choose an external environment model that is more stressing than a proposed system can tolerate.

2. Most wireless communications simulations (e.g., OPNET) lack adequate representations for multipath fading and distortion, and for jamming other than broadband noise jamming. These effects can often be much more important for overall system performance than background noise. Broadband spread spectrum confers partial immunity against multipath and narrowband jamming (e.g., swept tones) that can totally disable a wireless system using conventional narrowband waveforms. However, the benefits of the (military-unique) broadband spread spectrum waveforms are not seen when one makes comparisons using models that do not include these effects. Support for military use of commercial narrowband waveforms is undoubtedly based in part on grossly optimistic assessments of performance that do not reflect realistic military operating environments.

3. Channel models are often inextricably interwoven with the system model, i.e., elements of the system (modulation, demodulation, and other analog signal processing) are lumped together with the channel to form a single "discrete" channel model, a black box whose inputs and outputs are symbols (bits or groups of bits). The primary disadvantage of such models is that one cannot separate the system from the channel in order to compare different systems against the same channel. See Appendix C for a brief overview of the different types of channel models.

4. Even when there is a clear separation between the communications system and the channel in the model, the coding of the interface between the two may not be clean and, in any case, varies from one simulation to the next. This prevents one from easily removing the channel part of a simulation in order to substitute a different channel (or removing the system part in order to substitute a different system). One of the DARPA GloMo (Global Mobile) working groups on wireless communications[49] is currently developing a simulation application programming interface (API) standard.

[49]Led by Professors Rajive Bagrodia and Mario Gerla of UCLA.

The goal of being able to make fair comparisons of competing systems using existing (i.e., future) simulations without extensive recoding will not be realized until all four of the above problems have been solved.

6. Recommendations for 6.1/6.2 Funding

Note: 6.1 and 6.2 funding refer to government program lines for basic research and advanced development, respectively. The appearance of these numbers as section numbers below is coincidental.

6.1 The Problem of Duplicative Research

Funding by the military and various government agencies is often duplicative. A certain amount of duplication may actually be desirable, especially when different research groups take different approaches to the same problem. There is, however, excessive research effort in some areas and not enough in others. There are at least two possible factors that contribute to this situation. One is a lack of coordination among funding agencies. The other is that some research areas are simply more glamorous than others; research officers (and the higher-level decision makers above them) are naturally more interested in funding "hot" topics, or at least topics that have some name recognition, and tend to shy away from topics where there is currently little activity. However, it is often these unpopular, unexplored areas where research dollars have the greatest potential to uncover something new.

6.2 Batteries for Portable Electronics: An Area Where Funding Should Be Reduced or Eliminated

In our opinion, there are some areas where government funding is being used for research that the private sector is actively conducting on its own, and where government funding may be unnecessary. A prime example is research on batteries. Because of the large and growing portable electronics market, the private sector has ample incentive to develop batteries with higher energy densities and longer lives. Research in zinc-air batteries (see Chin, 1997) and a number of other new battery technologies is being carried out, much of it without government involvement. The argument that the military has a unique need for batteries that can operate at low temperatures does not seem valid, since outdoor usage of portable electronic equipment in various industries exposes equipment to low temperatures and other harsh conditions.

The plastic lithium ion battery (not to be confused with the conventional liquid lithium ion battery used in notebook computers) was developed with military funding (Sewell, 1996). This battery does have some features that are attractive for military applications, including greater safety (some types of high energy density batteries can catch on fire when punctured) and the ability to mold the battery into virtually any shape. However, there are also commercial battery applications with similar requirements. For example, safe high energy density batteries are important for future automotive applications because of the potential for battery ruptures in a collision.

We believe that developments in low power electronics will greatly reduce the power required to do advanced signal processing. Of course, one factor that limits the maximum transmission range is transmitted EIRP, which in turn depends on transmitted power. Higher battery energy density would permit increased transmitter output power; in practice, however, low antenna heights are the key limitation for transmission ranges between hand-held and/or manpack equipments. In any case, batteries are only a concern for dismounted soldier networks. Much of the Army's interest in mobile wireless networks is for vehicle-mounted radios, where power is not a big concern. For the hierarchical networks discussed in Sections 3.3 and 3.4, the nodes that require large transmitter output powers are the mobile base stations; these would almost certainly be powered by either diesel generators or fuel cells rather than batteries. Thus, in our opinion:

Higher battery energy densities are not a prerequisite for the development of future mobile wireless military networks.

6.3 Recommended Funding Areas and Topics

6.3.1 Channel and Interference Modeling

Continuous Channel Models

In order to enable fair comparisons of competing wireless systems via simulation, there is a need for libraries of validated channel models for fading channels and other time-varying channels. Needed are moderately fast-running impulse-type ("continuous") channel models for different combinations of terrain type, transmitter-receiver separation, antenna heights, transmitter/receiver speeds, and carrier frequency. The internals of the models, as well as the data and the methodology by which the models are fitted to the data, should be available to any researcher who wants to independently check any step of the process by which the model was created. Possible specific research topics include:

A. Plans for how tests should be conducted in order to gather the necessary data.

B. Frequency-independent methods for representing both flat fading and frequency-selective fading channel characteristics.

C. Suitable data structures for storing the model parameters, as well as fast algorithms for accessing the database and for interpolating in carrier frequency, time, and other variables.

Statistical Modeling of Interference

Better methods for statistical modeling of co-channel and adjacent-channel interference sources in the aggregate, including:

A. Better statistical tools for checking whether measured interference data agrees with a given time series model, or for selecting the model which best fits given data. The emphasis here should be on practical statistical tools for stationary non-Gaussian time series, because adequate tools for Gaussian time series already exist. By "practical" we mean tools that (1) are applicable to classes of nonGaussian time series models that are potentially useful for characterization of wireless interference, and (2) lead to algorithms with modest computational requirements (fast-running algorithms). Efficient simulation of nonGaussian time series models might also be a fruitful area of research; however, we note that this is an area which has already received a fair amount of attention.

B. Libraries of validated statistical interference models for different combinations of environment (urban office, urban highway, urban light industrial, suburban residential, etc.) and frequency. Again, general availability of the data, at least to qualified academic researchers for purposes of doing independent validation, is important.

Predicting Mutual Interference via Simulation

There is a need for methods for simulating mutual interference effects in mobile wireless networks via finite element and sampled waveform techniques with variable (selectable) accuracy. The goal is to be able to predict bit error rates and packet error rates due to multiple interference sources, including terrain masking, specular multipath, diffuse multipath, and diffraction. A possible subtopic involves the development of techniques for rapidly assessing which of a large number of interference sources will contribute significantly to the in-band interference power seen by a given receiver, so that the number of interference sources for which high-accuracy propagation calculations must be performed can be kept to a minimum.

6.3.2 Components and Subsystems

Broadband Devices and Amplifiers

There seems to be a lot of work in the area of low-power electronics (e.g., the DARPA Low-Power Electronics Initiative), but relatively little research on broadband devices, especially broadband (linear) high-power amplifiers at frequencies above Ka-band. The spectrum on the slopes of the 60 GHz oxygen absorption band is extremely valuable for military wireless communications. It is desirable to transmit at a frequency that simultaneously minimizes the risk of jamming and detection, while achieving the desired bit error rate at the intended receiver. One method for accomplishing this is to slide away from the peak absorption frequency until the desired link quality is achieved. Further developments in the area of broadband devices, and broadband high-power amplifiers for use at frequencies near 60 GHz in particular, are needed for high-capacity trunking (see the discussion of the hierarchical mobile network in Chapter 3).

Adaptive Notch Filters

Adaptive notch filters are useful for combating narrow-band interference and jamming in spread spectrum systems. They are particularly attractive for use in military variants of commercial spread spectrum receivers, since the addition of the notch filter has essentially no impact on the rest of the receiver design. Although research in digital notch filtering is still being done, we have included this topic under "components and subsystems" because the most pressing questions involve the choice of implementation; development of new filters does not appear to be necessary. Notch filters are already being used successfully to protect radars against unwanted narrow-band signals.

The simplest adaptive notch filter produces a single notch of fixed width, with only the on/off status and notch center frequency being adapted. A more sophisticated adaptive notch filter might adapt both the notch center frequency and the width, produce multiple notches, or both. The key questions to be answered for military wireless communications applications of adaptive notch filters are:

A. What is the tradeoff between complexity of the filter and system performance for scenarios involving small numbers of narrow-band jammers?

B. What is the best implementation approach that can be realized at low cost?

Passive Radiometers for Ground Terminals

Use of passive radiometers in ground terminals and line-of-sight radios is a practical but largely unexplored method for providing the operator with information about the status of the link. A passive radiometer can be used to detect increased system noise temperature by making power measurements at the edges of the signal band (this measurement would presumably be done by processing the Intermediate Frequency signal). Noise temperatures of 150 to 290 degrees Kelvin would indicate rain in the transmission path. A noise temperature of 8000 degrees K might be caused by the presence of the sun within the antenna field of view (e.g., for an air-to-ground link). A noise temperature of 80,000 degrees might be caused by a nuclear detonation or, more likely, jamming. A noise temperature of 800,000 degrees would indicate jamming (or faulty hardware). Possible research topics might include investigation of:

A. the operational use of antenna noise temperature information.

B. low-cost approaches for integrating radiometers into ground terminals and radios.

6.3.3 Signal Processing

Bandwidth Efficient Modulation and Coding

A. There is a need for additional progress on bandwidth efficient modulations that are at the same time reasonably power efficient and suitable for multiple carrier operation (FDMA), with or without power balancing. Current research on Gaussian Minimum Shift Keying (GMSK), especially in combination with turbo codes, looks very promising, but there is certainly more room for research here, e.g., to find good codes for various combinations of code rate and decoder delay, and to address the "decoded error rate floor" problem. Also, experimental hardware developments for purposes of demonstration and to permit testing in the laboratory against various channel degradations could help to move this technology more quickly into military applications.

B. Waveforms currently in use for military wireless, including ground, air, and air-to-ground links, have poor spectral efficiencies (e.g., roughly 0.02 bits per second per Hertz for the Joint Tactical Information Distribution System). There is a need for bandwidth efficient techniques for air-to-air and air-to-ground links, as well as for other links with slowly varying signal-to-noise ratios. The ability to rapidly and efficiently assess the link quality (e.g., by re-encoding and comparing against the demodulator output stream) and to switch to a lower or higher code rate (or to a different symbol rate or symbol alphabet) in order to achieve as much throughput as possible with changing channel conditions, while also achieving a minimum specified error rate, would greatly enhance performance over present systems.

Multiuser Detection for CDMA

One of the major drawbacks of Code Division Multiple Access (CDMA) spread spectrum for military applications is the so-called "near far" problem, in which interference from a nearby transmitter swamps the desired signal from a distant transmitter. The near-far problem tends to be worse for military CDMA systems than for commercial IS-95 cellular systems because, for a variety of reasons, it may not be practical to use power balancing, so that powers of received signals may differ by 30 dB or more. Multiuser detection techniques offer considerable promise for mitigating the near-far problem, but they require significant increases in complexity.[50] One could investigate receivers that make use of reduced-complexity multiuser detection techniques. Also of interest would be demonstrations of hardware implementations and small-scale experiments to evaluate performance.

Combining Signals from Multiple Receivers

Consider a hierarchical network of the type presented in Chapter 3. In such a network, spare trunk capacity might be available much of the time, depending on traffic patterns. An interesting use of spare trunk capacity is for enhanced processing of weak signals or signals affected by jamming. For such signals, rather than receiving and processing the signal at a single base station, one might employ two or three base stations. One of these base stations, which we will call the "primary base station," would receive raw demodulated bit streams (hard decisions or soft decisions) from the other base stations. Error control decoding would be performed using the bit streams from all available sources. Research questions include:

A. Appropriate combining methods for using multiple demodulated streams.

B. Simulation studies to estimate the potential benefit over conventional single-receiver processing.

6.3.4 Network Protocols

"Reliable UDP"

Research on routing for transmission of short messages over multihop wireless networks (packet radio network networks) has been ongoing for a number of years. Most of this work was funded by DARPA, but there has also been some Army-funded work, e.g., the recent research of Berry et al. (1997) on minimum-energy routing for wireless networks. The minimum-energy routing problem is interesting from a theoretical point of view; however, maximizing message delivery probability and minimizing delay are objectives that have greater practical value for the Army.

Most of the network literature on minimum-delay routing tends to assume reliable links, while literature on reliable delivery tends to ignore delay (and factors that delay depends on, such as capacity). An interesting and useful problem is the transmission of high priority (high precedence) messages over a wireless network in which links have various levels of reliability. Sending the same message via two disjoint routes is an approach that might be used to simultaneously

---

[50]Highly linear low noise amplifiers with large dynamic ranges (to avoid saturation) may be needed.

achieve high delivery probabilities and low delay, while also minimizing the impact on the rest of the (lower priority) network traffic. (Flood routing maximizes the probability of delivery, but may have unacceptable impact on other network traffic). Research objectives might include:

A. The development of a suitable metric or metrics that combine delivery probability and delay,

B. The development of efficient algorithms for choosing disjoint routes that maximize this metric given known link reliabilities, and

C. Techniques for estimating link and route delivery probabilities in a dynamic network.

Protocols for Combined Direct and Relayed Communications

The use of communications relays on UAVs is attractive for a number of reasons. However, one of the arguments against the use of a single UAV relay is that this creates a single point of failure for the network. An interesting area of investigation is the combined use of UAV relays (two-hop communications) with direct line-of-sight communications. Direct line-of-sight communications could be used as a backup in case of UAV failure, to reduce the load on the UAV, or to achieve Low Probability of Detection for short-range communications. Questions that might be studied include:

A. Algorithms/protocols for establishing links, including determining whether the UAV relay is to be used for a given connection.

B. Techniques for increasing system capacity, including directive antennas on the relay and/or on the mobiles, and output power control (in order to minimize interference to other users).

C. Evaluating the performance of such networks for various types of terrain and distributions of the mobiles.

Routing and Queue Management Algorithms for Handling Precedence and Perishability

The urgency of a message can in general be characterized by using precedence levels, perishability times, or both. The precedence level indicates importance, whereas the perishability time specifies when the information must be received in order to be of value. Most military messaging networks use precedence, but not perishability. However, tagging messages with a perishability time may be essential for wireless networks that are severely constrained both in link data rates and queue buffer space. A wide variety of network algorithms that handle precedence, but not perishability, already exist. Possible research topics include:

A. The development of queue management algorithms ("queue disciplines") that handle both precedence and perishability. For a queue in which packets (or messages) are characterized by precedence only, the most common queue discipline is head-of-line (HOL) priority queueing (see Kleinrock, 1976). With this discipline, higher-precedence packets jump ahead of all lower-precedence packets in the queue; packets having equal precedence are processed in order of arrival. However, when packets are characterized by both precedence and perishability time, the best strategy is not obvious. Clearly, a packet whose perishability time has already expired (or will expire before it can reach the head of the queue) should be removed from the queue. One might in some cases want to process a lower-precedence packet whose remaining lifetime is short ahead of a higher-precedence packet with a longer lifetime.

B. The development of routing algorithms that take both precedence level and perishability into account, as well as perhaps link delivery probabilities and mean incremental delays. (In some cases, one might choose to use a route that is less reliable but faster, or the converse.)

C. The problem of how to assign precedence levels and perishability times for different types of command and control, intelligence, and logistics messages. There are really two questions here: (1) What are the appropriate values to use for various types of messages (e.g., call for fire, routine status and position report, etc.)? This might depend on the nature of the operation. (2) What procedures or other mechanisms should be used to ensure that the precedence level and perishability time are correct?

7. Conclusions

7.1 Applicability of Commercial Technology, Products, and Services

To what degree can (or should) the Army make use of commercial standards and products for mobile wireless communications? The answer is somewhat different in each of the areas that we examined.

7.1.1 Components and Subsystems

For many types of components and subsystems, the higher commercial grades are as good as or better than their MILSPEC counterparts. However, in a few areas, such as antennas and broadband high-power amplifiers, the military has unique needs.

Where there is insufficient commercial demand for a class of component or subsystem that the military needs, it may be worth taking aggressive measures to ensure that a reasonable pool of both suppliers and R&D technical expertise continues to exist. Mulholland (1998) discusses an interesting case history involving suppliers of gallium arsenide semiconductors for the military.

The military needs to weigh the advantages of relatively inexpensive "disposable" equipment that is designed for shorter lifetimes. These advantages include reduced load on maintenance depots and a reduced time lag for the introduction of new technology. For those operations where high availability is essential, the most practical way to ensure availability might be for each company to carry an extra radio or two. Designing for long mean times between failure (MTBFs) under extreme conditions results in high costs that are hard to justify.

7.1.2 Waveforms and Signal Processing

In this area, there is a sharp divergence between military requirements and commercial practices.

7.1.3 Middle-layer Protocols (OSI layers 3–4)

The type of mobility support provided by mobile IP and IPv6 will not serve the Army's needs for mobile wireless networking. Although protocol features that allow applications and users to make the most of whatever network resources are available at any time is an urgent requirement for the military, commercial protocol standards are not moving toward the incorporation of such features. Multicast routing will be essential for future military wireless networks. However, because commercial multicast routing protocols are being designed for wired networks with static topologies and reliable links, they will not be useable in fully mobile wireless networks.

On the other hand, it will be difficult for the military to develop competing products because of the complexity of the protocols and the difficulties associated with testing. It is important for the military to remain engaged in the IP standards process and to fund research to develop middle-layer protocols that will be better suited for military mobile wireless networks.

7.1.4 Commercial Products and Services

Commercial wireless equipments are generally unsuitable for most Army operations. Use of wireless LANs, cordless phones, and other short-range systems at base and in rear areas are possible exceptions. In our opinion, tactical use of existing cellular telephone networks makes sense only for peace keeping and disaster relief operations in urban and suburban areas (assuming that the disaster has not rendered the system unusable).

Most commercial LEO and MEO satellite systems provide only low rate data services; the military needs higher data rates. For high data rate systems such as Teledesic, the terminals will not be fully mobile. High data rate fully mobile communications could be achieved by building military-unique terminals with phased array antennas for use with Teledesic; however, the costs of such terminals would be high. For all of these LEO satellite systems, financial viability of the service providers, as well as security and robustness against jamming, continue to be major concerns for all of these services.

Conclusions of this section are summarized in Figure 3:

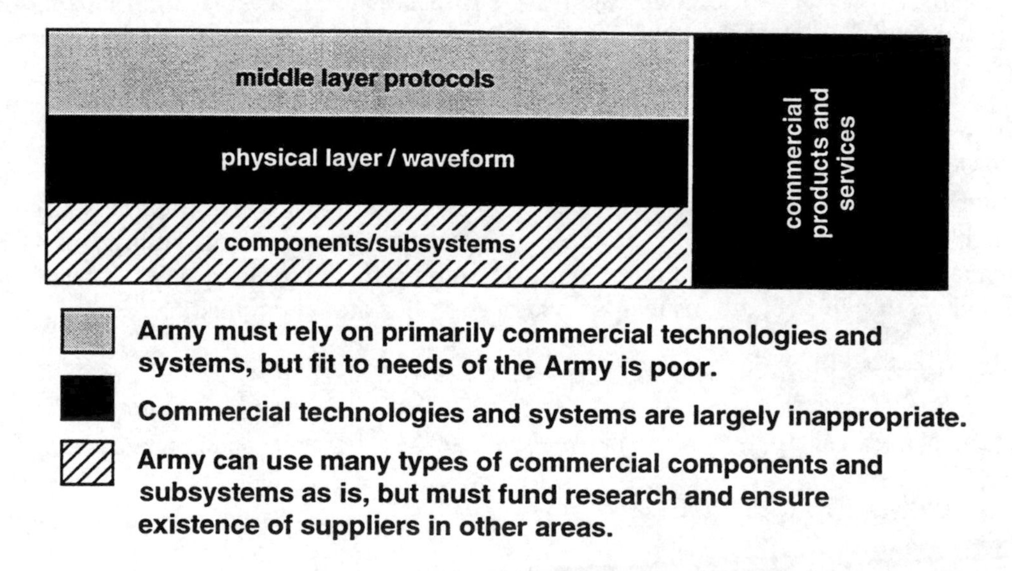

Figure 3—Which commercial technologies can be used on the tactical battlefield?

7.2 Two Key Findings of This Study

1. Existing and emerging commercial wireless standards are gradually addressing many of the communications problems that must be solved in order to meet the needs of the commercial world for wireless voice, wireless e-mail access, and related services. However, at both the physical layer and at higher layers, choices are being made that are fundamentally incompatible with Army tactical operations and with the Digital Battlefield concept. Although variants of commercial systems would appear to offer a middle course of action between military-unique and commercial off-the-shelf systems, this approach is typically not practical because costs quickly approach those of military-unique systems. One important exception

to this is ruggedized enclosures. A second is the addition of adaptive notch filters for rejection of narrowband interference.

Commercial wireless (terrestrial and space-based) systems and services will not meet the Army's future tactical needs, and the Army must consequently trade off requirements against future investments in research and Army-unique systems.

2. Military systems designers and planners have a critical need for simulation tools that can accurately predict the performance and behavior of mobile wireless networks operating in realistic tactical environments. Existing tools tend to concentrate on either the middle protocol layers or the lower "physical" layer, and do not simultaneously model both with sufficient detail and accuracy to yield useful results.

There is a need for (A) models that can accurately assess the impact of mutual interference (both co- and adjacent-channel) when large numbers of equipments operate in close proximity, (B) models that can be used to compare narrowband, frequency-hopping spread spectrum, and direct-sequence spread spectrum systems operating within a mobile network, including multipath effects, and (C) standard channel reference models against which competing system and network concepts can be tested.

Appendix A: A Brief Overview of Tactical Radio Communications

Current U.S. and NATO tactical ground, air, and air-to-ground communications depend on a potpourri of radios. Equipments currently fielded to ground forces alone (or at least in the inventory) include more than 40 types of radios, not counting variants. Some 200 radio procurements are currently "on the books," although it is expected that only a small fraction of these will actually be developed and fielded.

A few of the more widely used tactical radio families include:[51]

- The Joint Tactical Information Distribution System (JTIDS): a system designed for broadcast messaging and voice. All terminals have data capabilities; some have voice capability as well.
- The Single Channel Ground and Airborne Radio System (SINCGARS): a digital combat net radio with voice capability and limited data capability (1.2, 2.4, or 4.8 Kbps). The SINCGARS Improved Program boasts raw channel rates (i.e., before error control coding) of 9,600 bps and 16,000 bps.
- HAVE QUICK and HAVE QUICK II: a voice-only radio designed for point-to-point communications.
- Position Location and Reporting System (PLRS) and Enhanced PLRS (EPLRS), a data-only radio designed for point-to-point position reporting and messaging.

The history of military radios, in very general terms, divides into three major phases. In the first, pre-1970 phase, radios were analog and provided voice communications only. For many purposes, data transmissions are preferable to voice, and it is clear that the long-term trend in tactical communications is toward greater use of media other than voice, including strict-sense data, text messages, facsimile, and voice messages (store-and-forward voice); interactive voice communications is expected to shrink as a proportion of total traffic. Data offers many potential advantages over voice for tactical communications:

- Increased transmission reliability. A message or data packet lost due to noise or jamming can be automatically retransmitted. (Any digital information, whether voice or data, can be protected via error control coding, but interactive voice and other real-time transmissions are not amenable to retransmission).
- Reduced transmission times. Vocoded voice requires far more bits.[52] A text message requires far fewer bits than vocoded voice with the equivalent information content. A text message requires only a small fraction of the communications resources that would be required by an interactive voice message, and can also be delivered much more reliably than

---

[51]Watson (1992) gives a useful unclassified overview of military radios.

[52]A 600-character message can be read in roughly 30 seconds. With a 9600 bps vocoder (e.g., LPC-10), 30 seconds of speech corresponds to 36,000 bytes. At 7 bits per character, the 600-character text message corresponds to 525 bytes without compression. Note that the text message requires roughly 1/70th as much resources.

voice under adverse conditions, although there may not be a net time savings when time to compose the message is counted. Voice messaging represents an intermediate option between text messaging and interactive voice. Voice does offer some advantages over text messaging, including hands-free operation and speaker identification;[53] voice also conveys the speaker's emotional state, which may be important in some situations.

- Faster and more reliable message generation. Numerical information can be sent faster and more reliably via direct computer-to-computer communications; the advantages are greatest when this eliminates the need for reading a display or retyping of information. Measured data, e.g., from GPS receivers and laser range finders, can be automatically incorporated into transmissions.
- Reduced probability of detection. Data communications, because of their comparative brevity, tend to be more conducive to LPD (low probability of detection) than voice communications. LPD is critical for activities such as reconnaissance because it reduces the risks to forward spotters.

In the second phase of military radio developments, which covers the 1970s and the 1980s, digital data-only radios were introduced to support position reporting, IFF, and messaging. Digital modulation[54] offers many advantages over analog modulation; four of these are particularly important:

- The ability to send data.
- Encryption (of voice or data) is possible only with digital systems.[55]
- Error control coding, which increases the reliability and quality of transmissions over channels affected by noise, interference, or fading, is possible only for digital systems.
- With digital modulation, energy can be distributed in ways that hide the signal in order to provide for LPD or resistance to jamming (ECCM).

Phase-2 systems suffered from three major problems:

1. The necessity of using separate equipments for voice and data.[56]
2. Low data rates (typically 75 to 1,200 bps or less) for most of the data radios.
3. For most of the data radios, the waveforms and protocols limited them to point-to-point or netted line-of-sight communications, i.e., it was not possible to send a message to someone

---

[53]If a terminal fell into enemy hands, it would be easier to send a falsified text message than to hold a conversation, impersonating the rightful operator.

[54]Digital modulation means that information (voice or data) is transmitted as a sequence of discrete symbols, each of which represents a small number of bits. Voice must be converted to a digital stream, a process which is performed by a vocoder.

[55]Analog voice can be scrambled, but this is much less secure than encryption of digital voice and also tends to degrade intelligibility.

[56]JTIDS is technically an exception, since JTIDS also has a voice mode. However, in order to support voice, a large fraction of the throughput of a net must be used; it appears that operational use of the JTIDS voice mode is very limited.

unless a clear line of sight was available.[57] Some air-to-air links (e.g., JTIDS) were designed with a relay capability to support beyond-line-of-sight connections, but setting up a relay requires manual configuration by an operator, and the use of the relays also significantly degrades overall system performance.

In the third and most recent phase, radios have been developed that provide higher data rates, as well as radios that have both data and voice modes (with digital voice). For example, the Improved Data Modem (IDM)[58] supports data rates up to 16 Kbps. Note, however, that the combined voice/data radios in general do not support simultaneous data and voice. The ability to transmit simultaneous voice and data is an essential ingredient of the Digitized Battlefield concept, since this permits automatic status information to be transmitted as data together with speech, with no interference between the two, and without the need for an additional channel.[59] Thus, there is still a need for radios that integrate voice and data and that address problems 2 and 3 above.

This most recent phase also reflects the gradual recognition that it is inefficient to acquire a multiplicity of single-function radios; it is more cost-effective to acquire, maintain, and transport (as well as easier to train with) a smaller number of more general-purpose systems. The PLRS radio, a phase-2 system designed to perform only a single function—position reporting—was replaced by the more general-purpose EPLRS radio (which is still extremely limited in data rate).

Some operational concepts call for tactical transmission of high-resolution imagery, SAR data, and multimedia, including video. Such communications concepts will require data rates one to two orders of magnitude higher than those offered by existing radios. These fourth phase radios are still in the early stages of development.

---

[57]Excessive range, diffraction losses, or atmospheric absorption might prevent one from communicating even if line of sight were available.

[58]IDM is actually a modem that interfaces with any of several radios through an Intermediate Frequency input.

[59]Combined voice and data generally requires that the voice be converted into packets; however, packetized voice is a well-proven technology that is used, e.g., in Internet telephony. Once the voice has been packetized, the data can be inserted into breaks in speech (typical interactive speech is characterized by a voice activation factor of 40 percent, which means that 60 percent of the channel would be available for data). Alternatively, one might choose to use variable-rate voice coding, in which a control signal switches the vocoder to a lower rate whenever a data transmission is about to begin.

Appendix B: A Brief Review of Multiaccess Communications

B.1 Multiple-Access Communications

In a multiple-access system, users transmit information (e.g., voice or data) to one another using a shared communications medium. This shared medium could take any of several forms. Here are three examples of multiple access systems:

1. In a mobile satellite system with a single earth-coverage beam, ground terminals transmit to the satellite using one frequency band; the satellite rebroadcasts the signal using a different frequency band. All terminals (including the sender) hear the rebroadcast after a delay which includes the two-way propagation time and any processing delays.

2. In a Local Area Network (LAN), two or more computers are connected to a common bus (coaxial cable or fiber optic cable). If one computer transmits a message, all other computers hear the transmission after a delay.

3. In a packet radio network for mobile users (manpack radios or radios on ground vehicles), a user's transmission can be received by other users within some range which depends on the transmit power level and other factors.[60]

Although the shared medium is quite different in the above examples, a common feature of all three systems is that when any single user transmits, many other users can receive the transmission. In a mobile satellite system, users receive all transmissions from other users. A full-duplex radio can also monitor the rebroadcast of its own transmission. A half-duplex radio can do this only if the sum of the transmission duration and the *turn-around time* (time for the radio to switch from send mode to receive mode) is less than the round-trip delay. In the LAN, users receive all transmissions except their own. In the packet radio network, in general, each user can receive the transmissions of some subset of the user population (this subset is different for each receiver, and "A can hear B" does not imply "B can hear A"). For any of these systems, suppose that user A can receive transmissions from B and C, and that B and C transmit messages during overlapping time intervals. Consider first the case where signal strengths at the receiver are comparable. Unless strong forward error correction coding is being used, an overlap of only a few bits is sufficient to practically guarantee the loss of both messages, i.e., nothing intelligible is received; this event is called a *collision*. If signal strengths are sufficiently unbalanced, the more powerful of the two signals will be correctly received; this phenomenon is called *capture*.

A multiple-access protocol is an algorithm that coordinates user transmissions, including transmission of new packets and resolution of collisions (for protocols which permit collisions). Over the last few decades, a wide variety of multiple-access communications protocols have been proposed for different combinations of operating environment, patterns of usage, and user requirements. Why are there so many different multiple-access schemes? Part of the explanation is that

---

[60]Most military radios lack the capability for packet/burst transmission. However, newer radios will have this capability.

protocols suitable for some applications are often inadequate for other applications. Issues that must be considered in the selection or design of a multiple-access protocol include:

- What is the mean duration of a message transmission or of a call? In particular, the ratio of this duration to propagation delay is of fundamental importance.
- Do users generate messages or calls at fairly regular times, or in a bursty (irregular) fashion?
- What is the distribution of message length or of the call holding time? (Call holding times always have some variability; message lengths may be either fixed or variable).
- How critical is delay? In some applications, e.g., file transfer, only the time at which the last piece of a file is received is of interest, and even this may not be very critical. For packetized voice, packets must be delivered with low, nearly constant delay. For some message traffic, there is no hard limit on delay, but the value of the message may fall off rapidly with increasing delay.
- Are user equipments operating in a benign environment, or are high levels of noise, interference, or jamming present?
- To what extent can a message or call tolerate lost packets or corrupted bits?
- For systems that must support traffic of different types, e.g., voice, file transfers, and short data transmissions, the mix of these traffic types is generally important.

B.2 Basic Waveform Types

There are many possible ways to share a given band of frequencies among a population of users. The most basic waveform types are:

1. TDMA—Time division multiple access. Time is divided into slots, typically of fixed length. A given transmission must fall entirely within a single slot. In many TDMA systems, slots are grouped into frames as shown in Figure B.1. In TDMA with fixed assignments, the *k*th slot in every frame "belongs" to a given user, i.e., no other user may transmit in that slot.

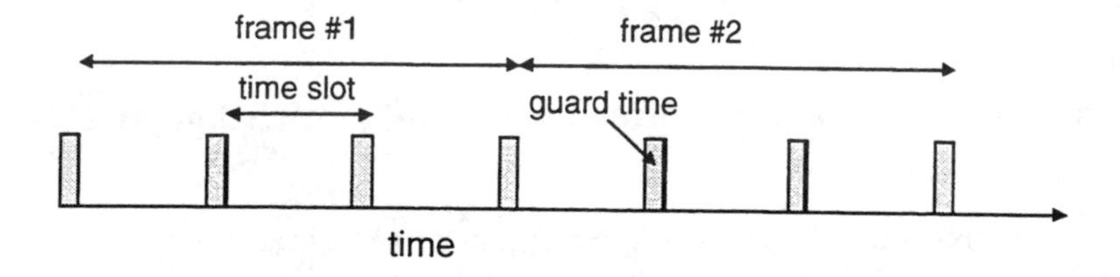

Figure B.1—Time division multiple access (TDMA) with frame structure

2. FDMA—The band is divided into smaller nonoverlapping frequency subbands, or channels, as shown in Figure B.2. A given transmission uses only one of these channels. With suitable guard bands between channels, simultaneous transmissions in different channels do not interfere with one another.

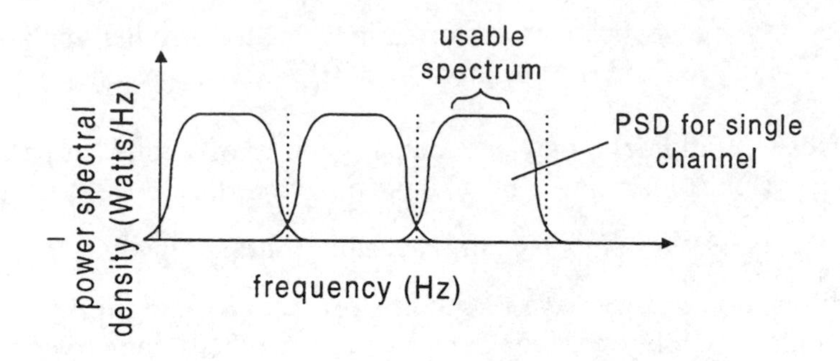

Figure B.2—Frequency division multiple access (FDMA)

3. FH SS—Frequency-hop spread spectrum. Both time and frequency are subdivided; a given transmission uses a pseudo-random sequence of transmit frequencies known as the *hopping pattern*. Gray squares in Figure B.3 correspond to a single hopping pattern. The receiver must be able to generate the same hopping pattern in synchrony with the transmitter. With *slow frequency hopping*, one or more symbols are transmitted on every hop. With *fast frequency hopping*, there are multiple hops per symbol. Slow frequency hopping is potentially more energy efficient because the receiver can use coherent demodulation. However, for low-data-rate military systems that use FH SS, fast frequency hopping is needed so that the hop rate is high enough to protect against follower jamming.

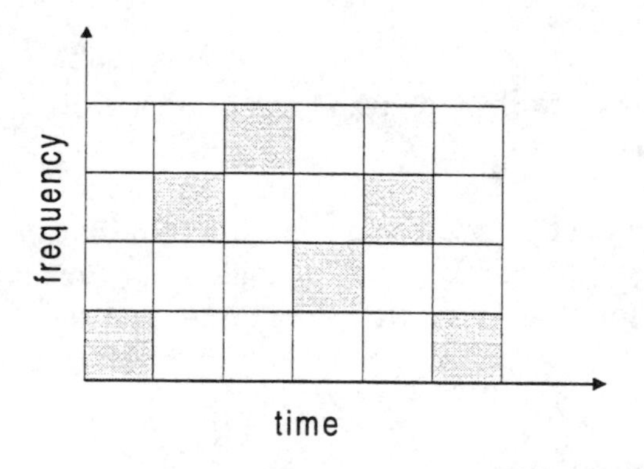

Figure B.3—Frequency hop spread spectrum (a single hopping sequence)

Although it is widely assumed that spread spectrum communications implies high cost, this is not necessarily the case. Hop rates up to 800 or 1,000 per second can be achieved using the frequency synthesizer in almost any military radio, i.e., a special fast frequency synthesizer is not required.[61]

4. DS SS—Direct-sequence spread spectrum. A bit stream can be viewed as a binary function, i.e., a continuous-time function that takes on the values 0 and 1. Consider a user data stream

---

[61]Cincinnati Electronics is providing a version of the Navy UHF WSC-3 radio to the Australian military for use as tactical line-of-sight radio. The Australian version, which has a hop rate of 800 per second, costs no more than the nonhopping version. (The U.S. Navy did not want the hopping version of the radio.)

having bit rate R_u. Prior to modulation, this stream is exclusive OR'd (XOR'd) with a pseudo-random (pseudo-noise) bit stream having rate $R_c >> R_u$. R_c is known as the *chip rate*. Timing is controlled so that user bit boundaries coincide with pseudo-noise bit boundaries; this ensures that the output of the XOR gate appears to be completely random, regardless of the behavior of the user data stream. The combined stream, each bit of which is called a *chip*, is modulated and transmitted. This sequence of operations is shown in Figure B.4. The resulting spread spectrum waveform has a bandwidth larger than that of the unspread waveform by a factor R_c/R_u. With DS SS, a given transmission uses the entire band of frequencies for the duration of the transmission, but it can nevertheless cause minimal interference to other users in the same band (and incur minimal interference from them).

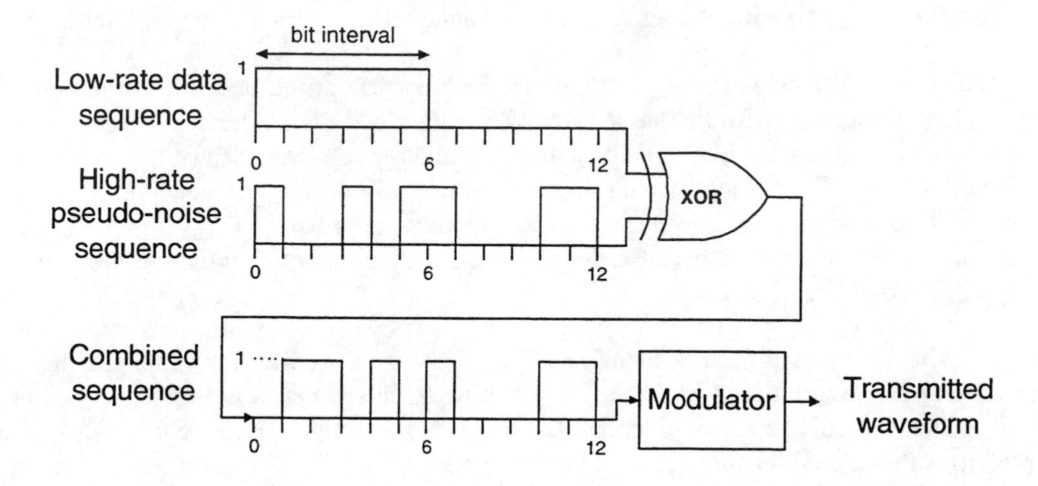

Figure B.4—A possible implementation for direct sequence spread spectrum (transmit side only)

5. Multicarrier DS SS. Multicarrier DS SS is similar to conventional DS SS, except that multiple carriers are used simultaneously by a single transmitter, with the same or different spreading codes on each of the carriers.

Note that a multiaccess scheme involves one or more basic waveform types and a multiaccess protocol that regulates the actions of users. For example, slotted Aloha is a multiaccess protocol for use in a single-channel TDMA system.

Various combinations of the basic waveform types are possible. Many wireless systems (e.g., the IS-134 digital cellular networks) use a combination of FDMA and TDMA; the total frequency band is first divided into multiple nonoverlapping channels, and each of these channels is then divided into fixed-length time slots.

B.3 Comparison of the Different Spread-spectrum Techniques

Typical hop rates for FH SS systems are on the order of 100 to 50,000 hops per second, while chip rates for DS SS systems can be several orders of magnitude higher. A consequence of this is that DS SS reception requires much more accurate synchronization, and synchronization is thus slower. Because of the comparatively short synchronization times, FH SS tends to be more suitable for burst transmissions, i.e., for short packet transmissions.

DS SS with conventional receivers requires very fine power control; otherwise, mutual interference between users severely limits capacity. Nonorthogonal FH requires some power control, although this can be much less precise. For DS SS with multiuser detection, which may become practical in the near future, power control requirements would be greatly relaxed.

With FH SS, one has the potential for considerable control over the shape of the power spectral density and can even make use of noncontiguous frequency blocks. This is not possible with conventional DS SS. Multicarrier DS SS does permit operation over noncontiguous frequency blocks, but it is easily implemented only when all frequency blocks are of equal width.

DS SS tends to be more power efficient than fast frequency hopping because of the noncoherent combining loss associated with the latter. Both DS SS and FH SS confer some immunity against multipath if the spread bandwidth is greater than the coherence bandwidth of the channel. Under optimal conditions, a DS SS system that has a rake reception with a larger number of taps can recover almost as much signal power as if no multipath were present. Conventional hopped systems do not use rake reception because the channel impulse response must be "relearned" each time the carrier frequency hops.

FH SS tends to be more effective than DS SS against narrow-band jamming and partial-band noise jamming, which are the most important jamming threats because high jammer power levels are easily generated only at relatively small fractional bandwidths. With DS SS, all symbols are corrupted to some degree by the jamming power, whereas with FH SS, a fraction of the symbols absorb the brunt of the jamming. For reasons that are complex, the latter situation results in error patterns that are more easily corrected by FEC decoders.

For further discussion of spread spectrum, we refer the reader to Magill et al. (1994) and to the text by Ziemer and Peterson (1985).

Appendix C: Types of Channel Models

A variety of types of channel models are in use. We consider here only strict-sense channel models, i.e., channel models that include propagation effects but not interference. Advantages and disadvantages of the three major categories of channel models are presented in the sections that follow.

C.1 Discrete Channel Models

Note that the word "discrete" refers to the values of the model input and output, rather than time, which is discrete for both discrete and continuous channel models. Discrete channel models lump the modulator and demodulator, and possibly also FEC coding, together with physical channel. Inputs to such a model are transition probabilities for symbols. These probabilities are in general time-varying and involve memory. Such models operate on sequences of bits or symbols, introducing errors to replicate the statistical behavior of the real demodulator output under stationary conditions.

The primary advantage of the discrete channel models is fast running times, i.e., computational requirements are minimal. On the other hand, with such a channel model it is impossible to separate the radio part of the model (hardware and signal processing algorithms used for modulation and demodulation) from the propagation channel part of the model, i.e., these two components of the physical layer become inextricably tangled. This is perhaps a good reason for excluding this category of channel model, since it is impossible to use them to make comparisons between different systems.

A special case of the discrete channel models is the discrete memoryless channel, or DMC. The DMC model is popular because it is trivial to implement in software, and it is important because, for most modulation formats, it is an exact description of the error statistics at the demodulator output when the signal is corrupted by Gaussian noise only (no multipath, narrowband interference, or other effects). For a DMC, each output symbol depends only on the corresponding input symbol (and the random noise), i.e., past input symbols have no effect, and the bit/symbol error rate (probability of error) depends only on the type of modulation and on the signal-to-noise ratio (SNR). For any modulation format, the so-called "waterfall curve" that translates from SNR to BER/SER can be obtained by calculation, by simulation, or by laboratory measurements.[62]

The DMC model is used in virtually all network simulation models that process bit streams (e.g., OPNET) because of its simplicity and fast running times. A disadvantage of the DMC, however, is that it produces independent bit and packet errors (often called "random" bit errors), a type of behavior that is unrealistic for many types of links. "To model mobile networks accurately, simulators . . . need to model the nature of errors on the wireless link precisely because errors are not uniformly distributed [independent] but rather tend to cluster" (NRC1, 1997, p. 85). In order to product "bursty" (clumped) errors, a channel model must have memory.

---

[62]Measured and calculated waterfall curves tend to agree closely, but may be slightly different because of implementation losses.

Discrete channel models with memory can be represented mathematically using an extension of discrete-time Markov chains known as hidden Markov models (HMMs). Although such models are highly flexible, and can in theory be used to approximate virtually any type of stationary communications channel, there has been little practical application of these models in communications. The primary reason for this is that procedures for obtaining HMM parameter values require either difficult mathematical calculations or costly numerical calculations. In the latter case, HMM parameters are fitted to data (derived from either laboratory measurements or link-level simulations) using an iterative procedure that adjusts the parameter values until an objective function is minimized. Because the objective function surface often has many minima, finding a global minimum tends to be difficult. HMM channel models are brittle in the sense that virtually any change in the system (a change in received SNR, in transmitter-receiver separation, which affects SNR, in the channel impulse response, in the modulation format, or in the receiver signal processing) requires that new data be generated and the HMM parameters re-estimated. Furthermore, because the objective function has many minima, a small change in system parameters may cause relative heights of minima to change so that the location of the global minimum changes radically. Thus, one cannot in general interpolate in HMM model parameters to account for changes in system parameters, i.e., there is no way to avoid continual re-estimation of model parameters. Because of these deficiencies, the use of HMMs is unattractive for communications applications (although not for applications in speech processing and other fields).

C.2 Continuous Channel Models

These models reproduce the effects of the channel propagation via a channel impulse response; in general, this impulse response must be time varying in order to account for both stochastic variation and changes in geometry (e.g., the distance from transmitter to receiver). A brief overview of continuous models for wireless channels can be found in Rappaport (1996).

With such models, a cleaner separation between the radio and channel models is possible; modulation and demodulation are no longer part of the channel model (antennas are still lumped with the channel). A given simulation must use a continuous channel model that has been sampled at the appropriate channel symbol rate; if values of the impulse response corresponding to a different time step are available, one can generally interpolate to produce values at the desired time step.

In order to handle mobility, one must recompute the channel impulse response whenever conditions change enough to make a significant difference. (How the recomputation times would be determined is unclear.) In order to handle different carrier frequencies, terrain types, and movement rates, one needs a family of channel models. The user would presumably make discrete choices to select among these, although carrier frequency might be allowed to vary continuously if a sufficiently accurate interpolation were possible.

C.3 Ray-tracing Models

Computational requirements of ray-tracing models are such that they will not be integrated into network-level simulations in the near future. Ray-tracing models are, however, important because they hold out the hope for accurate off-line calculation of impulse responses for specified transmitter-receiver geometries, terrain types, and antennas. Once these impulse responses have been generated, they can be used in a moderately fast-running network model.

Appendix D: Derivation of the Spectral Efficiency Form of Shannon's Capacity Formula

Our starting point is the standard form of Claude Shannon's formula for the capacity of a band-limited channel with additive white Gaussian noise (AWGN):

$$C = W \log_2\left(1 + \frac{P}{N}\right) \cdot 1\,\text{bit}, \tag{1}$$

where C is the capacity, or maximum average rate at which information can be transmitted over the channel, and has units of bits per second; W is the bandwidth of the channel in Hertz; and P/N is the ratio of the signal power divided by the noise power passed by the receiver front-end filtering (a dimensionless quantity).

In order to get a capacity equation involving spectral efficiency in terms of E_b/N_0, start by making the substitution $N = W \cdot N_0$ in (1). Manipulating, we get

$$\frac{P}{N_0} = W\left[2^{C/W} - 1\right]. \tag{2}$$

Dividing both sides of (2) by C gives

$$\frac{P}{N_0 C} = \frac{W}{C}\left[2^{C/W} - 1\right]. \tag{3}$$

To introduce E_b/N_0, we now reason as follows. When operating at capacity, the average energy per information bit equals the average signal power divided by the average information rate in bits per second, i.e.,

$$E_b = P/C. \tag{4}$$

Substituting in (3) using (4) gives a useful formula relating the achievable spectral efficiency C/W to the E_b/N_0 signal-to-noise ratio:

$$\frac{E_b}{N_0} = \frac{W}{C}\left[2^{C/W} - 1\right]. \tag{5}$$

Suppose we want to find the minimum E_b/N_0 required to achieve a spectral efficiency, C/W, of 6 bits/sec/Hertz. Substituting in (5), we find that the minimum $E_b/N_0 = 10.5 = 10.2$ dB. (To obtain the maximum achievable spectral efficiency for given E_b/N_0, one must solve numerically.)

References

Alwan, A., Bagrodia, R, Bambos, N., Gerla, M., Kleinrock, L., Short, J., and Villasenor, J., "Adaptive Mobile Multimedia Networks," *IEEE Personal Communications*, April 1996.

ASB, *Technical Information Architecture*, Army Science Board Summer Study, 1994.

Bateman, R., and Graff, C., "The CECOM Radio Access Point (RAP) Providing Integrated Voice, Video, and Data Service for the Warfighter," *Proceedings of the 1996 IEEE Military Communications Conference.*, McLean, Virginia, October 21–24, 1996.

Baum, C. W., *Decision-Theoretic Techniques for the Development and Use of Side Information in Frequency-Hop Radio Receivers*, doctoral thesis, University of Illinois at Urbana-Champaign, 1992.

Berry, R., Finn, S., Gallager, R., Kassab, H., and Mills, J., "Local Wireless Military Networks," *Proceedings of the Advanced Telecommunications/Information Distribution Research Program Annual Conference*, University of Maryland, January 21–22, 1997.

Bertsekas, D., and Gallager, R., *Data Networks*, 2nd edition, Prentice Hall, 1992.

Blahut, R. E., *The Theory and Practice of Error Control Codes*, Addison-Wesley, 1984.

Bojarski, J., "HCTR System White Paper," Version 4, U.S. Army CECOM Bulletin Board, 26 November 1996.

CECOM, *High Capacity Trunk Radio*, presented at the Warfighter's Information Network (WIN) Conference, May 20, 1996.

Chin, S., "Rechargeable Zinc-Air Batteries Vie for Portable Market: Thin Cell Combines Long Run Time and Small Size to Meet Leptop Computer Needs," *Electronic Products,* August 1997, pp. 17–18.

Corson, S., and Macker, J., *Architectural Considerations for Mobile Mesh Networking*, IETF Network Working Group, May 1996.

DoD, "Joint Technical Architecture," Version 5.0, Sept. 11, 1997.

Feldman, P. M., *An Overview and Comparison of DAMA Concepts for Satellite Communications Networks*, Santa Monica, California, RAND, MR-762-AF, May 1996.

Fiebig, U. C., *Spread Spectrum Techniques*, Institut für Nachrichtentechnik Report, 1998. This document can be found at URL http://www.op.dlr.de/ne/nt/NT-S/spreading_190198.html.

Frank, C. D., and Pursley, M. B., "Concatenated Coding for Frequency-Hop Spread Spectrum with Partial-Band Interference," *IEEE Transactions on Communications*, Vol. 44, No. 3, March 1996, pp. 377–387.

Gass, J. H., and Pursley, M. B., "A Comparison of Slow-Frequency Hop and Direct-Sequence Spread-Spectrum Systems for Different Multipath Delay Profiles," *Proceedings of the 1997 Military Communications Conference (MILCOM '97)*, Monterey, CA., November 1997.

Godara, L., "Application of Antenna Arrays to Mobile Communications, Part II: Beam-Forming and Direction-of-Arrival Considerations," *Proceedings of the IEEE*, Vol. 85, No. 8, August 1997.

Gordon, 1998, see http://www.gordon.army.mil/dcd/pss/hclos.htm.

Jane's Military Communications, 1998.

Johnson, D. B., and Maltz, D. A., "Protocols for Adaptive Wireless and Mobile Networking," *IEEE Personal Communications*, February 1996.

Kahn, R., Gronemeyer, S., Burchfiel, J., and Kunzelman, R., "Advances in Packet Radio Technology," *Proceedings of the IEEE,* November 1978.

Keller, J., "Armed Forces Set Sights on Tactical Networks," *Military and Aerospace Electronics,* January 1996.

Kleinrock, L., *Queueing Systems, Volume 1: Theory*, Wiley, 1975.

Kleinrock, L., *Queueing Systems, Volume 2: Computer Applications*, Wiley, 1976.

Leiner, B. M., Ruth, R. J., and Sastry, A. R., "Goals and Challenges of the DARPA GloMo Program," *IEEE Personal Communications*, December 1996.

Leiner, B. M., Nielson, D. L. , and Tobagi, F. A. (eds.), *Special Issue on Packet Radio Networks, Proceedings of the IEEE,* January 1987.

Lin, S., and Costello, D. J., *Error Control Coding*, Prentice-Hall, 1983.

Magill, D. T., Natali, F. D., and Edwards, G. P., "Spread-Spectrum for Commercial Applications," *Proceedings of the IEEE*, Vol. 82, No. 4, April 1994.

McEliece, R. J., *The Theory of Information and Coding*, Addison-Wesley, 1977.

MIT Lincoln Lab, *Architecture and Concept of Operations for a Warfighters Internet*, MIT Lincoln Lab for DARPA Information Systems Office, September 1, 1997.

Mulholland, D., "DoD Effort Lifts Gallium Arsenide Production Base," *Defense News,* May 4–10, 1998, p. 3.

NRC1, National Research Council, *The Evolution of Untethered Communications,* National Academy Press, 1997.

NRC2, National Research Council, *Energy-Efficient Technologies for the Dismounted Soldier,* National Academy Press, 1997.

NRC3, National Research Council, *Commercial Multimedia Technologies for Twenty-First Century Army Battlefields,* National Academy Press, 1995.

Office of the Assistant Secretary of Defense (OASD) (C3I) and Joint Staff Directorate for C4 (J6), *C4ISR Mission Assessment (CMA) Communications Mix Study* (briefing), 1997.

Pursley, M. B., *The Derivation and Use of Side Information in Frequency-Hop Spread Spectrum Communications*, IEICE Transactions on Communications, Vol. E76-B, No. 8, August 1993.

Pursley, M. B., and Wilkins, C. S., "An Investigation of Relationships Between Side Information and Information Rate in Slow-Frequency Hop Communications," *Proceedings of the 1997 IEEE Military Communications Conference (MILCOM '97)*, Monterey, CA., November 1997.

Rappaport, T. S., *Wireless Communications*, IEEE Press and Prentice Hall, 1996.

Rhea, J., "When Is a Component 'Commercial'?" *Military and Aerospace Electronics*, October 1996.

Rhea, J., "Digital Battlefield: Designers Still Have Much to Learn," *Military and Aerospace Electronics*, June 1997.

Ruppe, R., Griswald, S., and Martin, S., "Near Term Digital Radio (NTDR) System," *Proceedings of the 1997 IEEE Military Communications Conference (MILCOM '97)*, Monterey, CA., November 1997.

Sass, P., "Battlefield Information Transmission System (BITS) Far Term Strategy," Version 2.0, September 1997.

Sewell, K., "New Military Battery Technologies Trade Chemicals for Plastic," *Military and Aerospace Electronics*, December 1996.

Simon, M. K., Omura, J. K., Scholtz, R. A., and Levitt, B. K., *Spread Spectrum Communications Handbook,* Revised Edition, McGraw-Hill, 1994.

Tennenhouse, D. L., and Bose, V. G., "The Spectrum Ware Approach to Wireless Signal Processing," *Wireless Networks*, J. C. Baltzer AG, Vol. 2, 1996, pp. 1–12.

Torrieri, D., "Frequency Hopping and Future Army Wireless Communications," *Proceedings of the Advanced Telecommunications/Information Distribution Research Program (ATIRP) Conference*, University of Maryland, January 21–22, 1997.

U.S. Army, *Forward Support Battalion*, FM (field manual) 63-20, Chapter 4, 26 February 1990. Document can be found at URL http://www.atsc-army.org/cgi-bin/atdl.dll/fm/63-20/Ch4.htm.

U.S. Army Signal School, *WIN Master Plan*, Version 3, Directorate of Combat Developments, Fort Gordon, GA., June 1997.

Watson, A., *Future Trends in Radio Services Spanning the 2 MHz to 20 GHz Frequency Range for Different Geographic Regions,* Aerospace Report No. ATR-92(8182)-1, March 1992.

Ziemer, R. E., and Peterson, R. L., *Digital Communications and Spread Spectrum Systems*, Macmillan, 1985.